Volume 64

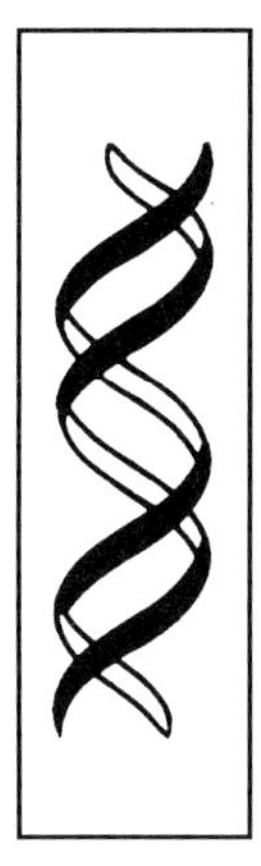

Advances in Genetics

Advances in Genetics, Volume 64

Serial Editors

Volume 64

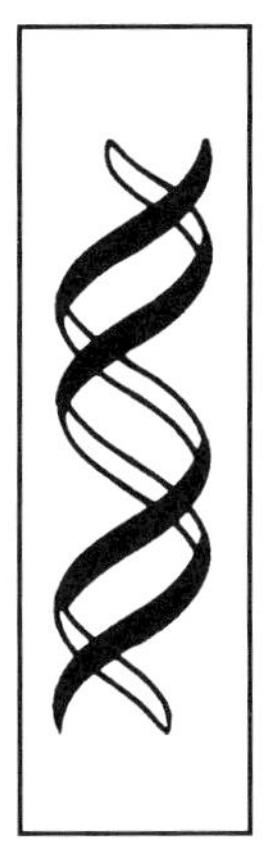

Advances in Genetics

Edited by

Jeffrey C. Hall
School of Biology and Ecology
University of Maine
Orono, ME, USA

Jay C. Dunlap
Dartmouth Medical School
Hanover, NH, USA

Theodore Friedmann
University of California at San Diego
School of Medicine
LaJolla, CA, USA

AMSTERDAM • BOSTON • HEIDELBERG • LONDON
NEW YORK • OXFORD • PARIS • SAN DIEGO
SAN FRANCISCO • SINGAPORE • SYDNEY • TOKYO
Academic Press is an imprint of Elsevier

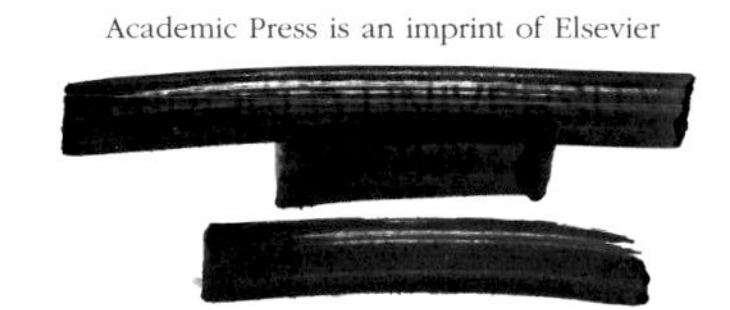

Academic Press is an imprint of Elsevier

525 B Street, Suite 1900, San Diego, CA 92101-4495, USA
30 Corporate Drive, Suite 400, Burlington, MA 01803, USA
32 Jamestown Road, London, NW1 7BY, UK
Radarweg 29, POBox 211, 1000 AE Amsterdam, The Netherlands

First edition 2008

ISBN: 978-0-12-374621-4
ISSN: 0065-2660

For information on all Academic Press publications
visit our website at elsevierdirect.com

Printed and bound in USA

08 09 10 11 12 10 9 8 7 6 5 4 3 2 1

Contents

CONTRIBUTORS

Numbers in parentheses indicate the pages on which the authors' contributions begin.

Diana Conte Camerino (79) Division of Pharmacology, Department of Pharmacobiology, Faculty of Pharmacy, University of Bari, I-70125 Bari, Italy

Xiao-Guang Chen (19) Department of Parasitology, School of Public Health and Tropical Medicine, Southern Medical University, Guang Zhou, GD 510515, People's Republic of China

Jean-François Desaphy (79) Division of Pharmacology, Department of Pharmacobiology, Faculty of Pharmacy, University of Bari, I-70125 Bari, Italy

Anthony A. James (19) Department of Microbiology and Molecular Genetics, University of California, Irvine, California 92697-4025; Department of Molecular Biology and Biochemistry, University of California, Irvine, California 92697-3900, USA

Marc Jeanpierre (1) Institut Cochin, Université Paris Descartes, CNRS (UMR 8104), Paris, France; Inserm, U567, Paris, France; AP-HP, hôpital Cochin, service de Génétique, Paris, France

Antonella Liantonio (79) Division of Pharmacology, Department of Pharmacobiology, Faculty of Pharmacy, University of Bari, I-70125 Bari, Italy

Geetika Mathur (19) Department of Molecular Biology and Biochemistry, University of California, Irvine, California 92697-3900, USA

Beston F. Nore (51) Department of Laboratory Medicine, Clinical Research Center, Karolinska Institutet, Karolinska University Hospital Huddinge, SE-141 86 Huddinge, Sweden

Csaba Ortutay (51) Institute of Medical Technology, FI-33014 University of Tampere, Finland

Sabata Pierno (79) Division of Pharmacology, Department of Pharmacobiology, Faculty of Pharmacy, University of Bari, I-70125 Bari, Italy

C.I. Edvard Smith (51) Department of Laboratory Medicine, Clinical Research Center, Karolinska Institutet, Karolinska University Hospital Huddinge, SE-141 86 Huddinge, Sweden

Domenico Tricarico (79) Division of Pharmacology, Department of Pharmacobiology, Faculty of Pharmacy, University of Bari, I-70125 Bari, Italy

Mauno Vihinen (51) Tampere University Hospital, FI-33520 Tampere, Finland; Institute of Medical Technology, FI-33014 University of Tampere, Finland

1

The Inspection Paradox and Whole-Genome Analysis

Marc Jeanpierre[*,†,‡]
[*]Institut Cochin, Université Paris Descartes, CNRS (UMR 8104), Paris, France
[†]Inserm, U567, Paris, France
[‡]AP-HP, hôpital Cochin, service de Génétique, Paris, France

Advances in Genetics, Vol. 64

0065-2660/08 $35.00
DOI: 10.1016/S0065-2660(08)00801-8

ABSTRACT

One of the major challenges of modern biology is distinguishing meaningful patterns from the random fluctuations of DNA sequences resulting from chromosome shuffling in each generation. A disease-causing mutation is more likely to be found in a large recombination interval. The paradoxical observation that causal genetic variants are more likely to be found in larger intervals is a consequence of sampling bias and is known as the inspection paradox. According to this paradox, the interval containing a fixed point (the causal gene variant) is around double the length of an interval not subject to this constraint, but this average doubling of length is attenuated or neutralized at the ends of chromosomes, where the distribution of interval sizes gradually returns to normal. This prediction is experimentally testable. The consequences of sampling biases for haplotype patterns are small in large studies of many families, but may be more marked when trying to counsel an individual family, because the doubling of the size of segments is only a large-number average, and the effect may be much larger for an unusual number of recombination events. The challenge of identifying a causal signature from haplotype patterns is illustrated by the problem of the proportion of X-linked mutations in pairs of affected brothers.

I. INSPECTION PARADOX

A. Introduction

The length of an interval that must contain a point is statistically larger than might be expected based on average spacings. This paradox has broad implications and has been given several names, including that of the "inspection paradox." It is often illustrated by practical examples, such as the speed of a car on a freeway or the time spent waiting for a bus to arrive at a bus stop. For most examples, the intervals considered are on a time scale, whereas, in genetics, the intervals are chromosomes segments between meiotic crossovers.

Terwilliger *et al.* (1997) showed that true positive peaks are wider than false positive peaks in genome-wide scans, suggesting that peak width may be useful in linkage analysis and providing a theoretical basis for these observations. These authors leaved it to subsequent studies to explore ways of taking advantage of length-biased sampling in gene mapping. This interesting hypothesis has been challenged (Lander and Kruglyak, 1995; Siegmund, 2001; Visscher and Haley, 2001). If the intervals around causal mutations are longer than would be expected if they were unconstrained, then measuring the size of genomic segments might facilitate the location and identification of the allelic variations underlying the phenotypes used as selection criteria.

Are intervals containing a genetic marker much longer than random intervals? Is it worth taking this difference into account in genetic analysis? Is this paradox a disappointingly simple bias, a pet problem of mathematicians or a possible nightmare for genetic counselors?

I discuss here the foundations of this theory and its possible implications for genetic analysis. Genome-wide analysis is now feasible and is likely to become cheaper and more reliable in the near future (Maresso and Broeckel, 2008). Genetic intervals between meiotic recombination events may be measured in familial studies, usually on a small scale, with each family referred for genetic counseling constituting a unique problem. The only segments easy to size when familial studies cannot be carried out for practical reasons are the long chromosomal segments of homozygosity markers, probably corresponding to autozygosity. The size of homozygous segments in the children of consanguineous marriages is a good model for illustrating selection bias (Section II). The inspection paradox predicts that the interval around a mutation may be approximately double that anticipated if mutations were evenly spaced. It may therefore be interesting to analyze the consequences of doubling average interval size when inherited segments account for about half a chromosome, as in nuclear families (Section III).

B. The origin of this name

Sampling paradoxes have been given several names including length-biased sampling or the waiting-time paradox. However, the term "inspection paradox" is frequently used because it seems less technical than many of the other names. The waiting-time paradox is probably one of the best known examples of length-biased sampling: intuitively, one might expect a passenger who arrives at a bus stop at some arbitrary point in time to have to wait half the mean time between two buses, whereas, in reality, they are more likely to have to wait the entire mean time between two buses, if buses do not arrive at regular intervals.

Such examples may attract the attention of those usually intimidated by the notions of sampling bias and statistics. However, beyond illustrating everyday experiences, this paradox is also of pedagogical value. Its mechanism illustrates several key features of the paradox: the intervals must be variable, with random starting points on the time scale, and observations must be made from a fixed point. If buses always arrive on time and are regularly spaced, there is no waiting-time paradox, and the mean time to the next bus is as expected a fraction of the mean time between two buses. This paradox can be applied to genetic problems, by replacing time with distance along a chromosome and the fixed point by the variation at a single allele. However, the genetic context differs from the bus paradigm in that chromosomes have a finite length, and there are few events (meiotic recombination events) to be considered.

The larger distance between flanking crossovers of a disease locus is often considered as well-established fact (Boehnke, 1994; Lange *et al.*, 1985).

Unfortunately, "waiting time" also has a different connotation in the medical world. An interesting "waiting-time" paradox, unrelated to the mechanism discussed here, has been described in the field of cancer epidemiology, in which the patients waiting longest for treatment have been shown to be the most likely to survive (Crawford *et al.*, 2002). In this example, the waiting time for treatment is a confounding factor, influenced by disease severity, because consultants tend to ensure that the patients at highest risk are treated more rapidly.

An individual gene is more likely to be found in a gene-rich region than in a gene-poor region. This bias resembles family size sampling biases, according to which, the probability of finding a family in a registry increases with the size of the family (Davidov and Zelen, 2001). It is important here to distinguish between a simple sample-size effect—the gene being more likely to be in the largest segment simply because that segment contains the largest number of genes—from the consequences of a specific distribution, in which the probability of a gene lying in a particular interval are not directly proportional to the size of the interval.

This difference may appear unnecessarily sophisticated, but it is of critical importance, as the simplest hypothesis is of little practical value whereas the hypothesis of a specific, potentially calculable distribution may facilitate the localization of causal allelic variants (Section II.B).

C. Basic elements and common misconceptions

1. Specific distribution

The inspection paradox does not depend on a specific algebraic function. It requires only the random distribution of the intervals. It is often presented in the context of renewal theory, but is not dependent on any particular mathematical model. This paradox has been demonstrated in a classroom environment, using an ordinary six-sided dice to simulate random intervals (Glocker, 1978); Glocker also presented a very convincing graphical explanation for nonmathematicians.

2. Double size

The average size of interval required to observe the desired event is double the mean interval between events. A common misconception is that the average interval size required is double that for a given model—for example, an exponential or power function—but that this interval may be larger than two for some imaginary highly variable distribution. This is not the case, because the intervals separating the fixed point from the nearest events on either side are the mirror images of a normal interval. The average size of the biased interval is therefore at most twice the size of the normal interval, depending on the model (Taylor and Karlin, 1998).

If the intervals are of equal probability, as the number of moves on a board game using the number showing on the dice, the average number of moves over a specific space is modestly increased from 7/2 to 13/3 from

$$\frac{1^2 + 2^2 + 3^2 + 4^2 + 5^2 + 6^2}{1 + 2 + 3 + 4 + 5 + 6} = \frac{13}{3}.$$

3. End effects

Statistically, an interval around a fixed point, an allelic marker, is likely to be larger than the *first* or the *last* interval. This is important for genetic analyses, because the consequences of biased sampling can be attenuated by considering large genomic segments. The gradual attenuation of the inspection paradox towards the telomeres has little effect on ancient mutations or in large families, because the ancestral interval around the mutation is small. However, this decreasing of the effect of the paradox is more apparent in nuclear families. This telomeric attenuation could be used in the experimental testing of the effects of sampling biases.

D. Experimental evidence

Most genetics papers consider the inspection paradox from a theoretical viewpoint. The paradox is counterintuitive, but of possible importance: is there any experimental evidence for it?

One of the key findings in studies of homozygosity in consanguineous individuals with autosomal recessive diseases is that the longest homozygous segment was associated with the disease more frequently than expected (Woods *et al.*, 2006). This study did not directly test the hypothesis that disease-associated segments are larger than nonspecific segments because the mutation-containing homozygous segments were not analyzed separately as a function of distance from the end of the chromosome. Taking the telomeres into account would probably have provided even stronger statistical evidence. In the next section, we will deal with the question of quantitative analysis of homozygous segment sizes.

One way of demonstrating the reality of the paradox would be to look for correspondence between the size of linkage intervals and position on the chromosome. Interval size also depends on the number of individuals studied, so such a plot can provide only a very crude picture of linkage interval variation as a function of position on the chromosome; such a plot is shown in Fig. 1.1.

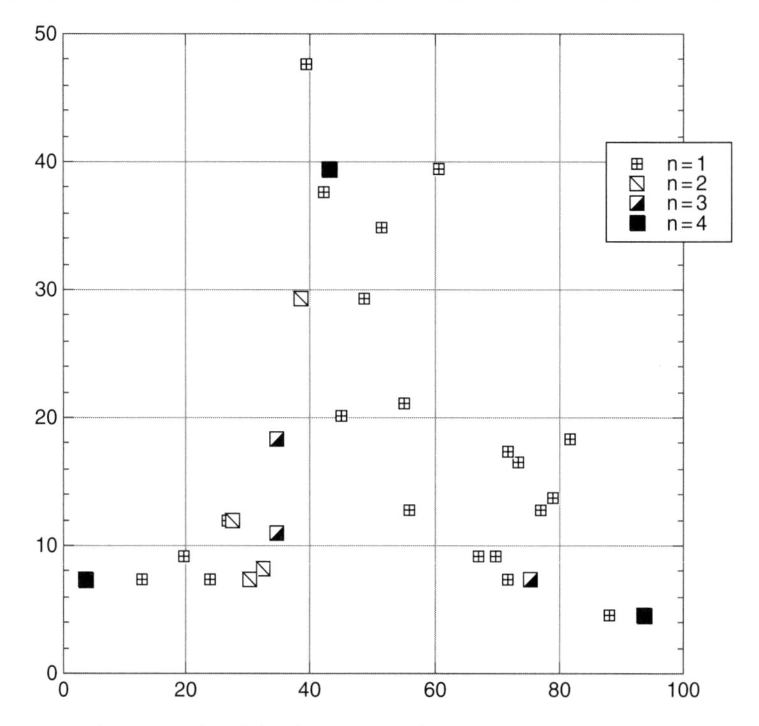

Figure 1.1. Linkage intervals and their location on a chromosome. Linkage intervals are plotted as a function of their midpoint (data source: Chiurazzi *et al.*, 2008). Sampling theory predicts a larger size for gene-containing segments toward the middle of chromosomes. *x*-axis: position (chromosome length = 100); *y*-axis: interval size.

II. HOMOZYGOSITY MAPPING

A. Simple models

Individuals with closely related parents have a higher risk than other individuals of recessive diseases. Searching the genome for segments shared by patients is a powerful empirical method for mapping disease genes (Houwen *et al.*, 1994). The offspring of consanguineous parents are likely to display disease-bearing homozygous segments, due to the inheritance of identical ancestral genomic segments from both parents, and these segments can be used for efficient homozygosity mapping.

The size of homozygous segments has been calculated by computer simulations of meiotic recombination. Clearly, all models are "wrong" in some respects, but models of meiotic recombination have several advantages over algebraic methods in that they are robust, because few rules are required to represent all familial links and inheritance. Highly realistic computer models can easily be constructed. The establishment of a mathematical model of recombination on a chromosome often requires simplifying assumptions, such as assuming the chromosome to be of infinite length. This is generally of little importance if there are a large number of meioses, because the ancestral region around the gene is small and likely to be some distance from the end of the chromosome. However, real chromosomes have ends.

This and the following section aim to provide a visual representation of genomic segments, facilitating graphical comparisons of biased and unbiased intervals. The theoretical description of the homozygous region from cousins is simple, and the density function is given by the power function:

> $(1 - \theta)^5$ and $(1 - \theta)^7$ for the child of first cousins and second cousins, respectively. The power function $(1 - \theta)^n$ describing the probability of an absence of recombination at a genetic distance θ from a mutation is usually approximated by an exponential function. As normalization constants, which are required to ensure that the probabilities add up to 1, are easier to calculate for power functions than for exponential functions, I will use power functions here (Hanein *et al.*, 2008).

When two children are considered, we can describe the homozygous regions shared by the two siblings as:

> $(1 - \theta)^7$ and $(1 - \theta)^9$ for a pair of siblings born to first cousins and second cousins, respectively.

Figure 1.2 shows that the size of a segment under the constraint of including a given point is larger than that without this constraint and could be represented by the function:

> $\theta(1 - \theta)^4$ for the offspring of first cousins and $\theta(1 - \theta)^6$ for the offspring of second cousins. These functions are good approximations for a marker in the central region of a real chromosome. The exact distribution is given by a combination of this function and a simpler power function. For telomeric genes, these expressions are truncated, according to the position of the mutation.

The ratio of the normal and biased distributions is therefore proportional to $(1 - \theta)/\theta$. The consequences of the selection bias are not linear and are particularly important for small intervals, for calculating the residual risk when aiming to exclude a disease.

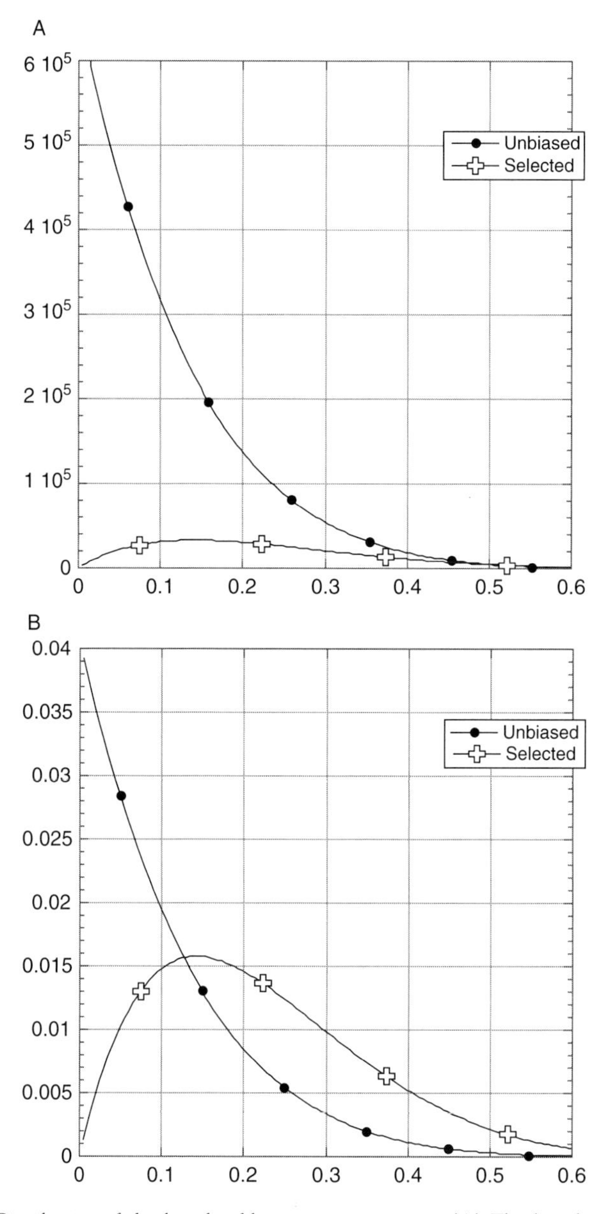

Figure 1.2. Distribution of the length of homozygous segments. (A) The length of homozygous segments is given by a power function (solid circle). The length of homozygous segments (second-cousin marriage) selected on the basis of containing a given allelic variant (cross) are statistically larger (y-axis: relative probability) than that of unbiased segments (solid circle). (B) The distributions depicted in (A) were normalized, ensuring

It is easy to show that the expected size of the interval is doubled by imposing the constraint of containing a fixed point, because, for an ordinary interval, the average size of a homozygous segment (offspring of first cousins) is 1/7:

$$\int_0^1 6\theta(1-\theta)^5 d\theta = \frac{1}{7}.$$

The first factor (6) is the normalizing constant making the integral of the density function equal to one. With the constraint of a fixed point, and a normalizing constant of 30, the average size is 2/7, as expected:

$$\int_0^1 30\theta^2(1-\theta)^4 d\theta = \frac{2}{7}.$$

A graphical representation of these distributions is shown in Fig. 1.2. The doubling of the average size of the region shared only by patients and not their unaffected siblings does not depend on the number affected or normal siblings.

B. Realistic models

The number of recombination events at each meiosis is not constant and may vary up to about five for an average human chromosome. Reconstructing haplotypes from familial studies is difficult or impossible with some oversimplified mathematical models. We must also take into account that long tracts of homozygosity are observed in normal individuals (Gibson *et al.*, 2006; Li *et al.*, 2006; Woods *et al.*, 2006).

Homozygous segments from individuals born to consanguineous parents are identical by descent, due to the inheritance of ancestral genomic segments from both parents. For offspring of first cousins, about half the chromosomes would be expected to contain at least one homozygous segment. The probability of observing several different homozygous blocks on a chromosome is distributed differently according to whether a geometric constraint is applied. The difference

that the sum of all probabilities was one. For homozygosity intervals of any observed length (*x*-axis), the probability of an allele-containing region is on average twice as large. Small homozygous segments are extremely unlikely to contain the gene, whereas the relative probability (selected/unselected) may be greater than two for larger segments.

is small, but may be important, because it suggests that the constraint of including a fixed point—variation at the causal allele—affects haplotype patterns. The "whole picture" provided by a chromosome is loosely defined, and the number of distinct homozygous blocks may be used as a surrogate for a "pattern."

The inspection paradox is well known to statisticians and its consequences are easily predictable. A key issue in terms of both practice and theory concerns the maximum amount of information that can be obtained from the genetic analysis of a family. In addition to the size of the interval around a marker, the number and position of meiotic recombination events may be important. The following example illustrates this hypothesis.

When a chromosome has two homozygous segments, one containing a mutation and another an "empty" block, the sizes of the two blocks are, paradoxically, not independent, because the size of the empty block affects the distribution of the mutation-containing segment (Fig. 1.3). This effect probably reflects the biased distribution of crossover events resulting from recombination.

As proximity to chromosome ends or telomeres also affects segment size, abstract model chromosomes of infinite size and with uniform recombination rates may not be the most appropriate models when high levels of accuracy are required. Thus, haplotype patterns are subject to several constraints, because whenever we study patients, the sample must contain at least one fixed point, the disease-causing mutation. The inspection paradox explains the reasons for which the variable interval continuing a mutation must be larger than an unconstrained interval. It is reasonable to assume that the skewed distribution of recombinant crossover around the allelic variant may affect other segment boundaries in models of chromosomes of a finite size. In the next section, I will illustrate the possible importance of the inspection paradox for genetic counseling.

III. X-LINKED MUTATIONS

A. Risk calculation and genetic counseling

Nonsyndromic autosomal recessive mental retardation is an extremely heterogeneous disorder. Most cases are unexplained and hundreds of genes may be responsible for mental retardation (Chiurazzi *et al.*, 2008). An extensive search would potentially be expensive, even with state-of-the-art oligonucleotide microarrays and the latest generation of high-throughput sequencing techniques. Furthermore, it may be difficult to establish a causal relationship between a missense mutation and the disease, and certain causal variations—intronic or extragenic variation in particular—are even more difficult to identify than missense mutations within reasonable amounts of time and at reasonable cost.

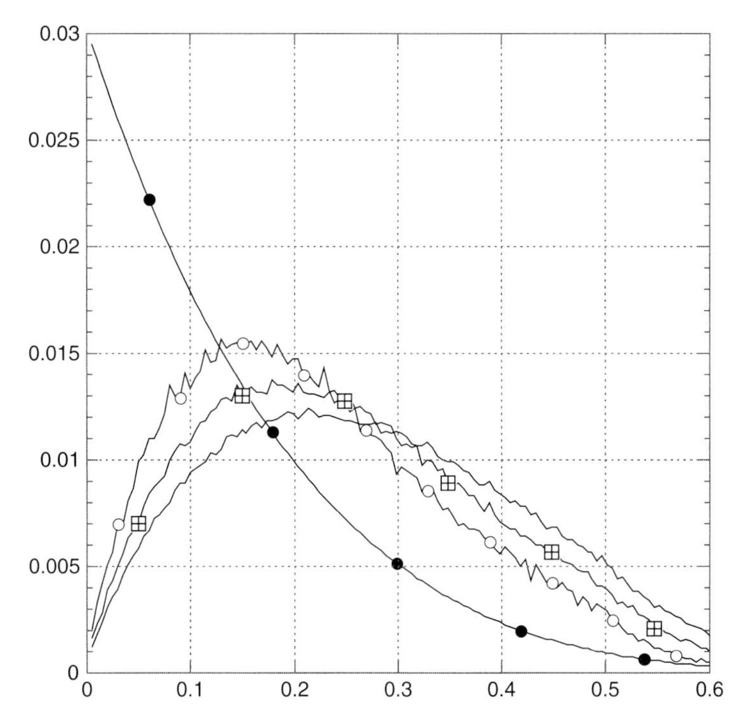

Figure 1.3. Importance of haplotype patterns. Two homozygous segments are observed on a chromosome. The sizes of the gene-containing and "empty" segments are represented (x-axis, fraction of chromosome length) by simulating meioses leading to the production of offspring first-cousin parents (y-axis, probability). The distribution of the mutation-containing segment is affected by the size of the empty segment. This suggests that it may be necessary to take into account the number and position of recombination events to extract maximal amounts of information from genome analysis. The figure shown is based on a small number of simulations of recombination.

With so many possible disease-causing genes and highly elusive mutations, indirect diagnosis may be a valuable tool for clinical geneticists. Indirect diagnosis requires a limited number of markers and provides a statistical estimation of risk within a known time window. The timing of result delivery can be scheduled, without requiring the sequencing of possible disease-causing genes. Psychologically, the risk estimates hardest to deal with are those in which uncertainty is close to the maximum, as for a one-to-one odds ratio. DNA analysis alters the prior probabilities and the residual risk may be very low, as shown in Fig. 1.4.

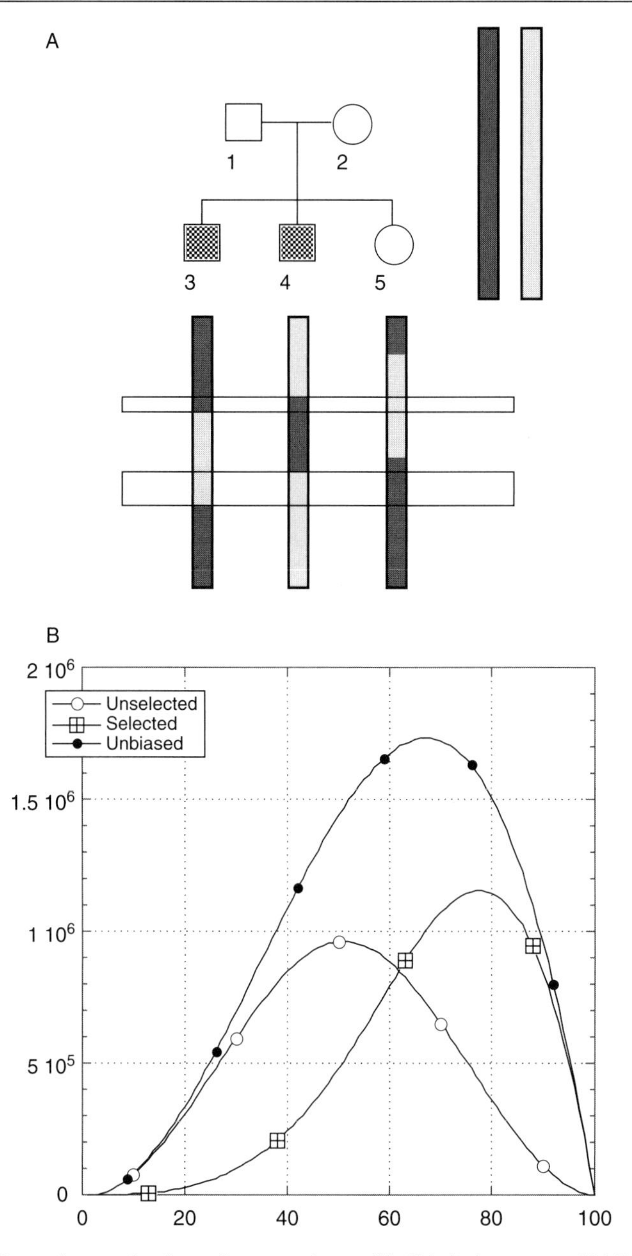

Figure 1.4. Interval size and indirect diagnosis of a possible X-linked mutation. (A) Two brothers with nonspecific mental retardation. Single-gene and complex inheritance are both possible and hundreds of genes may be responsible for the disease. An important issue for genetic counseling of the sister of these two affected brothers is the possibility of

The presence of two affected sons in a family may not necessarily be indicative of X-linked disease. This is because an excess of boys is commonly observed in malformative and genetic disease, even for mutations not located on the X chromosome. A large proportion of patients with nonspecific mental retardation have no precise diagnosis.

If mutation-bearing chromosomes and normal chromosomes differ in terms of their patterns of recombination, number and length of intervals, then it is possible to calculate the proportion of X-linked families. The pattern of mutation-bearing chromosomes is not intrinsically different; it is different because patients are *selected* on the basis of a phenotype, such as nonsyndromic mental retardation. The proportion of families with two affected boys could be calculated by considering the size of intervals and the number of recombination events. As most mutations on the X-chromosome are recent and correspond to different events, the effects of bias are easier to calculate, because the inspection paradox affects only once a sibship. This rather counterintuitive observation is explained by the specification of the problem, because the regions shared by two boys are also biased if the disease-causing gene is not on the X-chromosome (Fig. 1.4).

The genetic counseling problems most frequently encountered concern genetic diseases in small families, usually with one or two patients. What significance does doubling the size have when the inherited segments are about half the size of a chromosome?

B. Mutation location and haplotype patterns

The key regions are the genomic segments shared by the two affected sons. When the problem centers on the analysis of a small number of meioses, the actual number of meiotic recombination events should be preferred over the average distance on a model chromosome, because the topological constraints on haplotype reconstitution cannot be overcome with a predetermined fixed number of recombination events.

X-linked inheritance. Imaginary X-chromosomes are represented to illustrate the practical importance of indirect diagnosis. The residual risk may be calculated from the patterns expected in the presence or absence of a causal mutation. (B) The size of the mutation-containing segment is stochastically larger than that of segments not constrained to contain this mutation. The sum of the two segments shared by the brothers has been plotted (x-axis, length of the two shared segments) against relative probability (y-axis). The causal variant is assumed to locate close to the middle of the chromosome. Mutation-containing chromosomes (open cross) and unlinked chromosomes (solid circle) have different distributions. The third line (small cross) indicates the expected range of a general chromosome without selection. In this hypothetical situation, the prior probability of X-inheritance is assumed to be unknown. Similar theoretical distributions could be used to calculate the proportion of pairs of affected brothers with true X-chromosome mutations.

For a middle-sized chromosome about 188 cM in length, such as the X-chromosome, the actual number of recombinant crossover events varies from zero to about five (Tease *et al.*, 2002).

In the context of genetic counseling for a family with two affected sons, a modest relative risk (of about 2) may be obtained. These figures may appear limited and of little utility for a serious condition, but it should be borne in mind that taking all the information into account may results in significantly high or low figures. Severe X-linked mutations are recent, and have a half life of one generation. We have two variable parameters to calculate: the probability of the mutation being X-linked and, assuming this is the case, the position of the mutation on the chromosome.

Most of the families referred for genetic counseling are small, so every piece of information is potentially valuable for genetic counseling. For example, knowing that there are no patient with an X-linked disease in the ancestors and their descents in a large family, is not the same as having no information about ancestry, and this piece of information alone can halves the risk (Jeanpierre, 1988).

IV. DISCUSSION

A. A theoretical framework for "bad luck"

The negative aspect of the inspection paradox for gene localization is that the gene is found in a larger region than expected. Similarly, mapped genes are more frequently found in gene-rich regions than in regions with a low density of genes, often referred to as "gene deserts." This may provide a theoretical explanation for the "bad luck" of geneticists trying to find genes in large, difficult regions.

Terwilliger *et al.* (1997) made the interesting point that the length-biased or inspection paradox should not always be considered a nuisance because it can help to distinguish true peaks from false ones, based on the width of linkage scans. This thought-provoking hypothesis has been challenged by several authors (Lander and Kruglyak, 1995; Siegmund, 2001; Visscher and Haley, 2001). This chapter does not aim to provide a mathematical demonstration that the width of linkage peaks is an additional source of information, independent of peak height, but to present some of its experimentally testable consequences. According to the theory underpinning the inspection paradox, the first and last intervals are not constrained by a fixed point. This end attenuation or neutralization may be used as an indicator for analyses of the possible impact of the selection bias on experimental data. A rather crude test of the size effect is illustrated in Fig. 1.1, in which linkage intervals have been plotted as a function of position on the chromosome. By contrast to linkage analysis, in which every sample provides an independent element of information, the size effect of the inspection paradox is independent of sibship size.

The main issue in genetics in not the existence of biased sample, because all the basic elements of the paradox are clearly present: intervals of highly variable size and markers or mutations (fixed points) used for a selection. Instead, the chief concern relates to its possible consequences of practical importance for genome analyses:

- A quantitative approach is feasible. The ratio of the two distributions, selected/unselected, is approximately proportional to $\theta/(1-\theta)$. The ratio is not *exactly* proportional to $\theta/(1-\theta)$ when there are few meioses because of the telomere attenuation of the larger segments. The nonlinear relation originates from the density probability of the power function. In the simpler model of variable intervals of equal probability, as the results of an n-faced dice, the relation would be linear.
- As the patients are selected based on their phenotype, the assumption that a causal allelic variant is present may facilitate the identification of a mutation-bearing haplotype. It is widely accepted that the fixed point may be either a causal mutation, or selected arbitrarily (Siegmund, 2001; Terwilliger *et al.*, 1997), but most genome-wide genetic analyses are deliberately carried out on individuals with a particular phenotype, such as patients, requiring the presence of one or several of the allelic variants targeted by these studies.

B. Sampling bias and genetic counseling

The average doubling in the size of the interval studied is modest, but may have practical implications specific families. Many of the families referred for genetic counseling are small, and all relevant information should be taken into account.

As this doubling is an *average*, for some families, the likelihood of a mutation being in the at-risk interval may be three to five times higher. This variability is of little consequence in studies averaging over a large number of samples. However, geneticists are often trying to exclude a risk, rather than confirming "at-risk" status, and this requires a calculation of the residual risk. For the exclusion of a disease, the probability of a mutation being located in an at-risk residual region is not proportional to the length of this region, and may be neglected once calculated.

Genome-wide analyses are now feasible. Computer applications are required because "back-of-the-envelope" estimations of risk do not take the geometry of recombination events into account. Real chromosomes are not abstract lines of infinite length. The whole chromosome may be the meaningful unit of information, if every piece of information is valuable for genetic counseling.

C. Positive selection

Length-biased sampling theory explains why selecting individuals for the presence of a genetic marker makes it more likely that that marker will be found in a larger interval. A similar process may occur in a population in which physical traits are positively selected and, therefore, transmitted within a larger region. It would be difficult to test this hypothesis, because unexpectedly long haplotypes for a frequent mutation are thought to be a genetic signature of recent positive selection. Mutations associated with positive effects reproductive fitness rapidly reach a high frequency, and the young age of the variant may be reflected by the large region around the mutation. Therefore, if any direct selection of physical traits occurs—a model symmetric to the selection of patients from a population—its consequences for haplotype structure may be confounded with the size effect of a rapid expansion.

Single nucleotide polymorphisms (SNPs) potentially of great importance for function are more likely to be singleton SNPs than SNPs in intronic and intergenic regions (Ke *et al.*, 2008). Recombination hotspots are indirectly favored to mitigate the consequences of expansion for haplotype structures (Coop and Myers, 2007; Kauppi *et al.*, 2003). It remains unclear whether selection bias has an impact on haplotype patterns in populations.

D. Conclusion

Paradoxes are like double-edged swords, stimulating interest in some whilst confirming the conviction of others that statistical thinking is unnatural.

The costs of genotyping are falling, but the computational tools for identifying meaningful elements, and for diagnosis or gene mapping currently lag behind data acquisition. We need simple approaches for extracting the maximum amount of information from an apparently chaotic pattern of haplotype blocks.

Acknowledgment

I thank Julie Sappa for making useful suggestions.

References

Boehnke, M. (1994). Limits of resolution of genetic linkage studies: Implications for the positional cloning of human disease genes. *Am. J. Hum. Genet.* **55**(2), 379–390.

Chiurazzi, P., Schwartz, C. E., Gecz, J., and Neri, G. (2008). XLMR genes: Update 2007. *Eur. J. Hum. Genet.* **16**(4), 422–434.

Coop, G., and Myers, S. R. (2007). Live hot, die young: Transmission distortion in recombination hotspots. *PLoS Genet.* **3**(3), e35.

Crawford, S. C., Davis, J. A., Siddiqui, N. A., de Caestecker, L., Gillis, C. R., Hole, D., and Penney, G. (2002). The waiting time paradox: Population based retrospective study of treatment delay and survival of women with endometrial cancer in Scotland. *Br. Med. J.* **325**(7357), 196.

Davidov, O., and Zelen, M. (2001). Referent sampling, family history and relative risk: The role of length-biased sampling. *Biostatistics* **2**(2), 173–181.

Gibson, J., Morton, N. E., and Collins, A. (2006). Extended tracts of homozygosity in outbred human populations. *Hum. Mol. Genet.* **15**(5), 789–795.

Glocker, D. (1978). A graphic illustration of the waiting time "paradox". *Am. J. Phys.* **46**(2), 185.

Hanein, S., Perrault, I., Gerber, S., Delphin, N., Benezra, D., Shalev, S., Carmi, R., Feingold, J., Dufier, J. L., Munnich, A., Kaplan, J., Rozet, J. M., *et al.* (2008). Population history and infrequent mutations: How old is a rare mutation? GUCY2D as a worked example. *Eur. J. Hum. Genet.* **16** (1), 115–123.

Houwen, R. H., Baharloo, S., Blankenship, K., Raeymaekers, P., Juyn, J., Sandkuijl, L. A., and Freimer, N. B. (1994). Genome screening by searching for shared segments: Mapping a gene for benign recurrent intrahepatic cholestasis. *Nat. Genet.* **8**(4), 380–386.

Jeanpierre, M. (1988). A simple method for calculating risks before DNA analysis. *J. Med. Genet.* **25** (10), 663–668.

Kauppi, L., Sajantila, A., and Jeffreys, A. J. (2003). Recombination hotspots rather than population history dominate linkage disequilibrium in the MHC class II region. *Hum. Mol. Genet.* **12**(1), 33–40.

Ke, X., Taylor, M. S., and Cardon, L. R. (2008). Singleton SNPs in the human genome and implications for genome-wide association studies. *Eur. J. Hum. Genet.* **16**(4), 506–515.

Lander, E., and Kruglyak, L. (1995). Genetic dissection of complex traits: Guidelines for interpreting and reporting linkage results. *Nat. Genet.* **11**(3), 241–247.

Lange, K., Kunkel, L., Aldridge, J., and Latt, S. A. (1985). Accurate and superaccurate gene mapping. *Am. J. Hum. Genet.* **37**(5), 853–867.

Li, L. H., Ho, S. F., Chen, C. H., Wei, C. Y., Wong, W. C., Li, L. Y., Hung, S. I., Chung, W. H., Pan, W. H., Lee, M. T., Tsai, F. J., Chang, C. F., *et al.* (2006). Long contiguous stretches of homozygosity in the human genome. *Hum. Mutat.* **27**(11), 1115–1121.

Maresso, K., and Broeckel, U. (2008). Genotyping platforms for mass-throughput genotyping with SNPs, including human genome-wide scans. *Adv. Genet.* **60,** 107–139.

Siegmund, D. (2001). Is peak height sufficient? *Genet. Epidemiol.* **20**(4), 403–408.

Taylor, H. M., and Karlin, S. (1998). "An Introduction to Stochastic Modeling." Academic Press, New York.

Tease, C., Hartshorne, G. M., and Hulten, M. A. (2002). Patterns of meiotic recombination in human fetal oocytes. *Am. J. Hum. Genet.* **70**(6), 1469–1479.

Terwilliger, J. D., Shannon, W. D., Lathrop, G. M., Nolan, J. P., Goldin, L. R., Chase, G. A., and Weeks, D. E. (1997). True and false positive peaks in genomewide scans: Applications of length-biased sampling to linkage mapping. *Am. J. Hum. Genet.* **61**(2), 430–438.

Visscher, P., and Haley, C. (2001). True and false positive peaks in genomewide scans: The long and the short of it. *Genet. Epidemiol.* **20**(4), 409–414.

Woods, C. G., Cox, J., Springell, K., Hampshire, D. J., Mohamed, M. D., McKibbin, M., Stern, R., Raymond, F. L., Sandford, R., Malik Sharif, S., Karbani, G., Ahmed, M., *et al.* (2006). Quantification of homozygosity in consanguineous individuals with autosomal recessive disease. *Am. J. Hum. Genet.* **78**(5), 889–896.

2

Gene Expression Studies in Mosquitoes

Xiao-Guang Chen,*[,1] Geetika Mathur,†[,1] and Anthony A. James†[,‡]

*Department of Parasitology, School of Public Health and Tropical Medicine, Southern Medical University, Guang Zhou, GD 510515, People's Republic of China

†Department of Molecular Biology and Biochemistry, University of California, Irvine, California 92697-3900, USA

‡Department of Microbiology and Molecular Genetics, University of California, Irvine, California 92697-4025

ABSTRACT

Research on gene expression in mosquitoes is motivated by both basic and applied interests. Studies of genes involved in hematophagy, reproduction, olfaction, and immune responses reveal an exquisite confluence of biological adaptations that result in these highly-successful life forms. The requirement of

[1]These authors contributed equally to the manuscript

Advances in Genetics, Vol. 64
0065-2660/08 $35.00
DOI: 10.1016/S0065-2660(08)00802-X

female mosquitoes for a bloodmeal for propagation has been exploited by a wide diversity of viral, protozoan and metazoan pathogens as part of their life cycles. Identifying genes involved in host-seeking, blood feeding and digestion, reproduction, insecticide resistance and susceptibility/refractoriness to pathogen development is expected to provide the bases for the development of novel methods to control mosquito-borne diseases. Advances in mosquito transgenesis technologies, the availability of whole genome sequence information, mass sequencing and analyses of transcriptomes and RNAi techniques will assist development of these tools as well as deepen the understanding of the underlying genetic components for biological phenomena characteristic of these insect species.

I. INTRODUCTION

Mosquitoes are vectors of pathogens that cause serious human infectious diseases, such as malaria, dengue, and yellow fever. Increasing concerns with failures in existing control methods stimulate applied research, and genetic modifications of mosquitoes for population reduction and replacement are proposed as potential strategies to control mosquito-borne diseases (Curtis and Graves, 1988; Knols *et al.*, 2007). Basic molecular studies of mosquitoes reveal remarkable adaptations to facilitate hematophagy, reproduction, olfaction, and immune responses to pathogen challenge, as well as genetic bases for insecticide resistance (Attardo *et al.*, 2005; Enayati *et al.*, 2005; Meister *et al.*, 2004; Ranson and Hemingway, 2005; Ribeiro and Francischetti, 2003; Rützler and Zwiebel, 2005; Zwiebel and Takken, 2004). Both applied and basic science investigations benefit from advances in mosquito transgenesis technologies that include the discovery and development of transposable elements for germline integration of exogenous DNA, suitable marker genes such as fluorescent proteins, standardization of microinjection techniques, and characterization of promoters that drive tissue-, sex-, and stage-specific expression (Atkinson and James, 2002; Catteruccia *et al.*, 2000, 2005; Coates *et al.*, 1999; Conde *et al.*, 2000; Horn *et al.*, 2002; Jasinskiene *et al.*, 1998; Kim *et al.*, 2004; Kokoza *et al.*, 2001b; Lobo *et al.*, 2006). Furthermore, the availability of whole genome sequence information for three mosquito species, *Anopheles gambiae*, *Aedes aegypti*, and *Culex quinquefasciatus*, allows global characterization of sequence conservation and genome structure through comparative and functional analyses by which patterns of evolution in gene and protein families are detected (Holt *et al.*, 2002; Nene *et al.*, 2007; Waterhouse *et al.*, 2008). This information is complemented by mass sequencing and analyses of transcriptomes, which provide information on gene expression levels in whole animals or specific tissues. RNAi techniques circumvent difficulties of conventional mutational analyses in mosquitoes and permit detailed studies of gene function and regulation. Targeted knockdown of transcription products in specific tissues is used in conjunction with bioinformatic gene discovery approaches to validate results *in vivo*. These advances hold great promise for the development of novel tools for controlling pathogen transmission as well as reveal the underlying genetic components for a number of biological phenomena found in these highly successful insect species.

II. STUDY OF GENE EXPRESSION IN MOSQUITOES

A. Transgenesis

Transgenesis, the stable integration of exogenous DNA into the genome of a target organism, is a powerful tool for gene expression studies and has been achieved with several mosquito species, all of which are important vectors of human pathogens (Table 2.1). Mosquito transgenesis is based on the paradigm developed for the fruit fly, *Drosophila melanogaster*, in which DNA flanked by the inverted terminal repeat (ITR) sequences of a Class II transposable element is mobilized (excised from a donor plasmid and inserted into target DNA, most often the mosquito chromosomal DNA). Mobilization is catalyzed by the action of the corresponding transposase encoded on a separate (helper) plasmid and usually under control of inducible, *cis*-acting promoter DNA. Mobilization assays monitor the movement of a transposable element from one plasmid to another and are important for determining if an element has the capability to excise and integrate in the embryonic environment of a specific species (Atkinson and James, 2002). These studies are a useful prelude before committing significant effort and resources to developing an element for transformation. Transgenes

Table 2.1. Mosquito Species That Have Been Transformed

Mosquito species	Transposon	References
Aedes aegypti	*Hermes*[a]	Jasinskiene *et al.* (1998)
	Mos1 (mariner)[a,b]	Coates *et al.* (1998, 2000)
	piggyBac[a]	Kokoza *et al.* (2001a)
Ae. fluviatilis	*piggyBac*	Rodrigues *et al.* (2008)
Anopheles gambiae	*piggyBac*[a]	Grossman *et al.* (2001)
An. stephensi	*Minos*[a]	Catteruccia *et al.* (2000)
	piggyBac[a]	Nolan *et al.* (2002)
An. albimanus	*piggyBac*[a]	Perera *et al.* (2002)
Culex quinquefasciatus	*Hermes*	Allen *et al.* (2001)

[a]Mobility assay data published.
[b]Purified *Mos1* (*mariner*) transposase also successful.

may contain DNA of homologous, heterologous, or synthetic origin, and their expression properties depend on *cis*-acting elements linked to marker, reporter or effector genes.

A number of early attempts at mosquito transgenesis supported dedicated efforts to develop systems that were both repeatable and reliable (McGrane *et al.*, 1988; Miller *et al.*, 1987; Morris *et al.*, 1989). The transposable elements *Hermes*, *Minos*, *Mos1*, *piggyBac*, and *Tn5* can integrate DNA into mosquito genomes and provide the bases for useful transformation vectors (Table 2.2). These elements have different evolutionary histories with *Hermes* belonging to the *hAT* family, *Minos* and *Mos1* derived from the *mariner* superfamily, and *piggyBac* representing a family in which it is the prototype. *Tn5* is of bacterial origin and belongs to the *IS50* family of insertion sequences. Wimmer and colleagues (Berghammer *et al.*, 1999; Horn and Wimmer, 2000) developed a set of donor transformation plasmids based on some of these elements that contain a variety of marker genes and the corresponding helper plasmids. Transformation efficiencies are best measured as the number of independent integration events that occur per fertile adult (Adelman *et al.*, 2002) and generally vary from 1% to 10%. However, higher frequencies are reported with *piggyBac*, especially in *An. stephensi* (Adelman *et al.*, 2004; Kokoza *et al.*, 2001a; Nolan *et al.*, 2002). The elements integrate principally at their known consensus nucleotide sites (e.g., the dinucleotide, TA, for *mariner*-related elements), but no systematic studies of genomic integration sites are available, although it is reported anecdotally that they are random. *Hermes* has noncanonical integration behavior into chromosomes of *Ae. aegypti*, incorporating portions of the donor plasmid and deleting parts of the transgene, including the ITR sequences (Jasinskiene *et al.*, 2000). *Mos1* integrations into *Ae. aegypti* most often are single events while experiments using *piggyBac* in this species and *An. stephensi* often recover animals with multiple insertions per genome (Coates *et al.*, 1998; Kokoza *et al.*, 2001a).

The apparent random nature of integration makes transgenes subject to insertion site effects that result in unanticipated expression characteristics. The expression profiles of multiple independent insertions must be analyzed to determine expression properties intrinsic to the transgene construct from those imposed by the surrounding genome. The use of "insulator" sequences to mitigate insertion site effects has been adopted widely in *D. melanogaster*, but has yet to be proven robust in mosquitoes (Farkas and Udvardy, 1992; Gray and Coates, 2004; Kellum and Schedl, 1991; Sarkar *et al.*, 2006). Site-specific integration into the genomes could mitigate these effects if a "docking site" were located in a region of the genome free of surrounding influences (Morris *et al.*, 1991). Chimeric *Mos1* and *piggyBac* transposases were reported to result in site-directed integration in plasmid-based transposition assays in *Ae. aegypti* embryos (Maragathavally *et al.*, 2006). However, the sequence complexity of target plasmids is much lower than genomic DNA, and it is not clear how the observed

Table 2.2. Transposons Used in Mosquito Transgenesis

Transposon	Family	Origin	~Size (kb)[a]	ITR[b] length	Target site	References
Hermes	*hAT*	*Musca domestica*	2.7	17	GTNCAGAC	O'Brochta *et al.* (1996)
Mos1	*Tc1–mariner*	*Drosophila mauritiana*	1.3	~30	TA	Medhora *et al.* (1991)
Minos	*Tc1–mariner*	*Drosophila hydei*	1.4	254	TA	Franz and Savakis (1991) and Loukeris *et al.* (1995)
piggyBac	Novel	Lepidopteran baculovirus	2.5	13	TTAA	Fraser *et al.* (1996)
Tn5	*IS50*	Enteric bacteria	5.8	19	9 bp, variable	Berg *et al.* (1983)

[a]Approximate size in kilobases.
[b]Inverted terminal repeat length in nucleotides.

specificity extrapolates to mosquito chromosomes. The phage, ΦC*31*, was exploited to develop a high-efficiency, site-specific integration system for inserting exogenous DNA into *Ae. aegypti* (Nimmo *et al.*, 2006). Φ*C31* uses a self-encoded integrase to insert its ~43 kilobase (kb) genome into a specific site, *attB*, of the host chromosome (Groth and Calos, 2004). Integration is mediated by synapsing *attB* with a specific sequence, *attP*, in the phage genome. The resulting recombination event produces two new sites, *attL* and *attR*, that are not substrates for the integrase, resulting in stable integration of the Φ*C31* genome. Linking *attP* or *attB* to any circular DNA fragment can facilitate its stable uptake into an *attB* or *attP* site, respectively (Thyagarajan *et al.*, 2001). Once docking sites were introduced into the *Ae. aegypti* genome, subsequent integrations were 7.9-fold higher than the primary integration event (Nimmo *et al.*, 2006). This approach should allow the insertion of large fragments of DNA into the mosquito genome.

Postintegration stability of nonautonomous (lacking a source of transposase) transgenes has not been a problem in general in mosquitoes. While one construct based on *piggyBac* was noticeably unstable in *Ae. aegypti* (Adelman *et al.*, 2004), it has proven difficult to induce remobilization of most transgenes once they are inserted into the genome (Sethuraman *et al.*, 2007; Wilson *et al.*, 2003). The reasons for this include disruptions of the element structure upon integration (Jasinskiene *et al.*, 2000), but for those that inserted intact it is not known and likely is specific to the element. Engineered elements based on *Mos1* and containing a source of transposase (autonomous) were able to catalyze the chromosomal insertion of other elements carrying *Mos1* ITRs, indicating the transcription and translation of functional transposase, but no movement of inserted elements were observed (Adelman *et al.*, 2007). While most nonautonomous transposable elements are stable in the absence of the homologous transposase, it is possible that endogenous sources of transposase may destabilize them. Methods have been developed for postintegration elimination of all transposon sequences in fruit flies (Dafaalla *et al.*, 2006; Handler *et al.*, 2004), but these have yet to be applied to mosquitoes. However, experiments with *Cre*-mediated recombination of integrated DNA show that specific sequences bound by *loxP* sites can be excised at high frequencies from the *Ae. aegypti* genome (Jasinskiene *et al.*, 2003).

The first successful use of a marker gene to detect repeatable mosquito transformation used a wild-type copy of the *D. melanogaster cinnabar* gene to complement a white-eye phenotype in a mutant strain of *Ae. aegypti* (Jasinskiene *et al.*, 1998). However, the most robust and widely-applied marker genes are those developed by Wimmer and colleagues comprising coding sequences for enhanced green fluorescent protein (EGFP), the cyan fluorescent protein (CFP), and *Discosoma* sp. red fluorescent protein (DsRed) controlled by the *D. melanogaster Pax6* eye-specific enhancer–promoter combination (Berghammer *et al.*, 1999; Horn and Wimmer, 2000). Positive transgenic mosquitoes are screened easily by

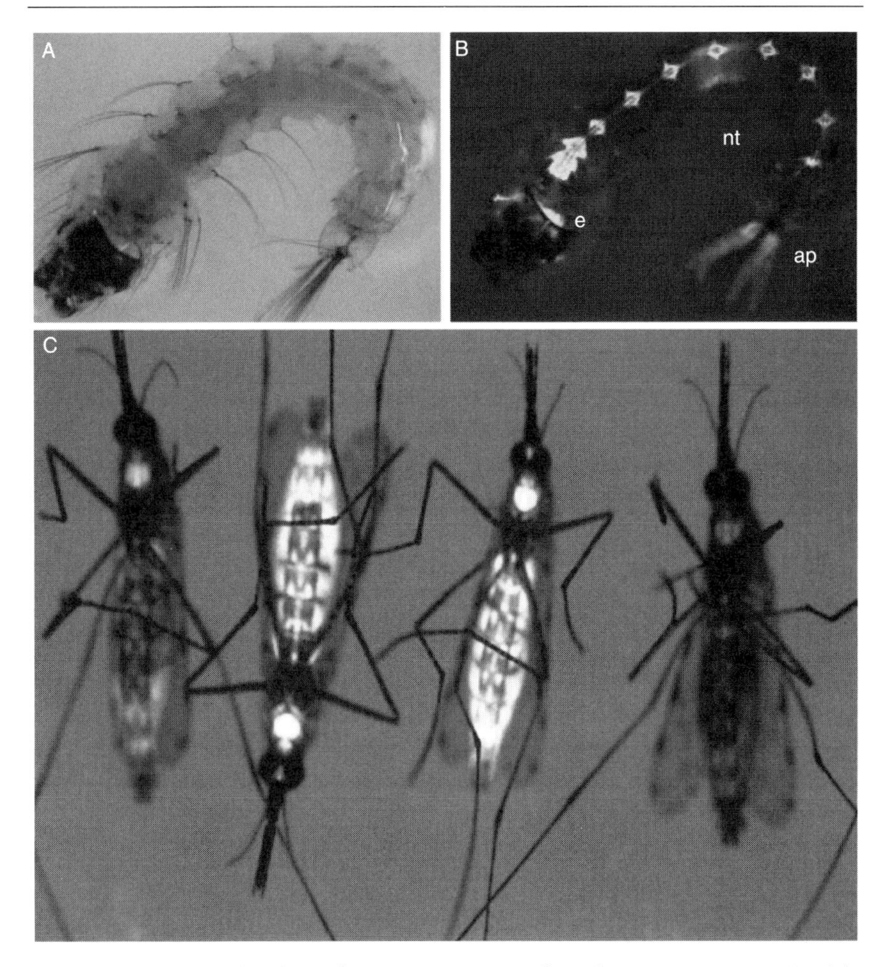

Figure 2.1. Expression of marker and reporter transgenes in the malaria vector mosquito, *Anopheles stephensi*. (A) Fourth-instar larva under ambient light. (B) Same larva as in (A) seen under fluorescence microscopy. Larval eyes (*e*), nervous tissue (*nt*), and anal papillae (*ap*) fluoresce with DsRed. (C) Adult females expressing the cyan fluorescent protein gene controlled by the *AsVg1* promoter (see Nirmala *et al.*, 2006). Specific fluorescence is detected in fat body tissues in the thorax and abdomen. Images courtesy of J. M. Sandoval.

observing the specific fluorescence in the eyes or other nervous tissue with UV-microscopy (Fig. 2.1). The anal papillae often fluoresce, but the reason for this is not know. Fluorescent marker genes behave as complete dominant alleles making it difficult to distinguish visually animals that carry one copy from those carrying two or more. Autofluorescence can confound the use of some coding sequences as reporter genes. Green and red autofluorescence are detected in the accessory glands of both transgenic and wild-type male *An. stephensi* (Catteruccia *et al.*, 2005) and a weak green signal was seen in the thorax of the wild-type mosquitoes (Yoshida and Watanabe, 2006). Detection of transgene expression products using immunoblots or gene amplification procedures for the transcription products may be needed to verify engineered expression (Chen *et al.*, 2007).

B. Evaluation of mosquito gene expression

The majority of reported gene expression analyses in mosquitoes use technologies developed in the last century and include Southern blots to determine gene copy numbers, Northern blot analyses to determine the size and presence of specific transcription products, hybridizations *in situ* to examine RNA distribution and accumulation in specific tissues, gene amplification for detecting the presence/absence and abundance of transcription products, and primary sequencing analyses of both genomic DNA and cDNAs for intron identification and processing of primary transcripts. A growing number of more recent studies have exploited genome projects and mass sequencing to look at genome-wide transcription profiles. These studies reveal intricacies of the evolution of these highly adapted ectoparasites as well form the bases for developing novel control strategies. Functional analyses remain critical for establishing the roles of specific DNA sequences in gene expression and modulation.

1. Analyses of mosquito-derived control DNA

Mosquito transgenesis procedures, although generally routine, are not trivial. As a result, a number of transient methods were developed for assaying promoter functions in whole animals or dissected tissues. Transient expression assays were developed prior to the availability of transgenesis methods to look at gene expression in *Ae. aegypti* salivary glands (Morris *et al.*, 1995). A liposome-based transfection reagent was used to introduce a DNA construct comprising the luciferase reporter gene under control of the *D. melanogaster* HSP70 promoter into cultured adult salivary glands. Luciferase activity was detected in glands indicating that, although terminally differentiated, the glands could take up and support the expression of exogenous DNA. A more detailed analysis was done on the γ-aminobutyric acid receptor (GABAR)-encoding gene (*Rdl*) of *Ae. aegypti* that confers high levels of resistance to cyclodienes (Shotkoski *et al.*, 1996). *Cis*-acting DNA containing the *Rdl* promoter was mapped to a 2.53 kb fragment following transient expression of plasmid constructs in microinjected embryos. A detailed study was made of the *Ae. aegypti* ferritin light-chain homologue (LCH) gene in cultured cells (Pham and Chavez, 2005). Transfection experiments indicate that this gene has a strong promoter, and DNase I footprinting identified a number of transcription factor-binding sites, including those that bind GATA, E2F, NIT2, TATA, and DPE. More recently, a transfection assay based on DNA microinjection into whole animals was used to analyze a vitellogenin-luciferase reporter gene in response to blood feeding (Isoe *et al.*, 2007). A small, 843 base pair (bp), fragment of the *Ae. aegypti vitellogenin-C* promoter directed a >200-fold tissue-specific induction. The size of the control DNA is consistent with prior observations in transgenesis studies with an *An. stephensi* vitellogenin-encoding gene promoter (Nirmala *et al.*, 2006). Functional mapping of the *Ae. aegypti* gene identified essential 5′-end regulatory elements in the region −780 to −182 bp from the transcriptional start site. Similar assays in the same study revealed a 1096 bp genomic fragment of the cecropin B gene could induce >100-fold expression in the fat body. These techniques need further development to be widely applicable for future gene analyses.

Mosquito genes of interest were defined first by their expression in tissues or body compartments in which significant interactions with pathogens were expected to take place (Meredith and James, 1990). Genes expressed in the midgut and salivary glands were isolated and characterized first. A number of efforts were made to assay mosquito *cis*-acting promoter DNA in *D. melanogaster* with variable success. A promoter fragment of the *An. gambiae D7r4* gene drove strong tissue-specific expression in the fruit fly, but only worked at a low level in *An. stephensi* (Lombardo *et al.*, 2005). Thus, transformation of the fruit fly is an unreliable assay for establishing expression properties of mosquito promoters. Following development of transgenesis, functional *cis*-acting control sequences were identified for genes encoding salivary gland secretory products, digestive enzymes, and proteins sequestered in the developing oocyte (Table 2.3).

Genes expressed in mosquito salivary glands encode proteins that facilitate feeding. While both sexes feed on sugars found in extrafloral nectaries, only adult female feed on blood and there are many genes whose expression is restricted to this sex. This latter category encodes a wide array of enzymes and other factors that counteract host hemostasis and immune system responses, and may initiate primary digestion of the blood (Ribeiro and Francischetti, 2003). For the most part, salivary gland gene expression appears constitutive, although some induction may be present after a blood meal to replenish stores (Marinotti *et al.*, 1990, 2005; Yoshida and Watanabe, 2006). Functional analyses of salivary gland gene promoters were stimulated because they were candidates for expressing antipathogen effector molecules. Genomic DNA fragments of the *Ae. aegypti Maltase*-like I (*MalI*) and *Apyrase* (*Apy*) genes were used to direct the expression of the luciferase reporter gene in transformed mosquitoes (Coates *et al.*, 1999). Both *MalI* and *Apy* promoter regions were capable of directing correct developmental-, sex-, and tissue-specific expression, but at low levels. In a separate study, an ~800 bp fragment of the *An. gambiae Apyrase* gene *AgApy* directed expression of a reporter gene in the salivary glands of transgenic *An. stephensi* (Lombardo *et al.*, 2005). Expression levels were low and the reporter gene product

Table 2.3. Functional Control DNA in Mosquitoes

Transgene		Expression profile[a]				
Promoter	Reporter	Species	Sex	Tissue	Temporal	References
Pax6	EGFP, DsRed	*Ae. aegypti, An. stephensi*	Male/female	Eyes	Constitutive	Catteruccia *et al.* (2005) and Nimmo *et al.* (2006)
AeVg	DefA	*Ae. aegypti*	Female	Fat body	Blood meal inducible	Kokoza *et al.* (2000, 2001b)
AeVg1	N2ScFv	*Ae. aegypti*	Female	Fat body	Blood meal inducible	Jasinskiene *et al.* (2007)
AeVgC	Luciferase	*Ae. aegypti*	Female	Fat body	Blood meal inducible	Isoe *et al.* (2007)
AgVgT2	EGFP	*An. stephensi*	Female	Fat body	Blood meal inducible	Chen *et al.* (2007)
AaVgR	DsRed	*Ae. aegypti*	Female	Ovary	Blood meal inducible	Cho *et al.* (2006)
AsVg1	CFP	*An. stephensi*	Female	Fat body	Blood meal inducible	Nirmala *et al.* (2006)
DmUb	N2ScFv	*Ae. aegypti*	ND[b]	ND[b]	Constitutive	Jasinskiene *et al.* (2007)
β2-tub	EGFP	*An. stephensi, Ae. aegypti*	Male	Gonads	Constitutive	Catteruccia *et al.* (2005) and Smith *et al.* (2007)
AeCP	Luciferase	*Ae. aegypti*	Female	Midgut	Blood meal inducible	Moreira *et al.* (2000)
AeCP	CecA	*An. gambiae*	Female	Midgut	Blood meal inducible	Kim *et al.* (2004)
AgCP	Luciferase	*Ae. aegypti*	Female	Midgut	Blood meal inducible	Moreira *et al.* (2000)
AgCP	SM1, PLA2	*An. stephensi*	Female	Midgut	Blood meal inducible	Ito *et al.* (2002) and Moreira *et al.* (2002)
AgAper1	Mutated PLA2	*Ae. fluviatilis*	Female	Midgut	Blood meal inducible	Rodrigues *et al.* (2008)
AgAper1	PLA2	*An. stephensi*	Female	Midgut	Blood meal inducible	Abraham *et al.* (2005)

act88F	EGFP	*Cx. quinque*	Male/female	Muscle	Constitutive	Allen and Christensen (2004)
actin	EGFP	*An. stephensi*, *Cx. quinque*	Male/female	Muscle	Constitutive	Allen *et al*. (2001) and Catteruccia *et al*. (2000, 2003)
actin5C	EGFP	*Ae. aegypti*	Male/female	All tissues	Constitutive	Pinkerton *et al*. (2000)
AAPP	DsRed	*An. stephensi*	Female	Salivary glands	Blood meal inducible	Yoshida and Watanabe (2006)
Apy	Luciferase	*Ae. aegypti*	Female	Salivary glands	Constitutive	Coates *et al*. (1999)
Mal1	Luciferase	*Ae. aegypti*	Female	Salivary glands	Constitutive	Coates *et al*. (1999)
Hsp70	Luciferase	*Ae. aegypti*	Female	Salivary glands	Adult	Morris *et al*. (1995)
Rdl	Luciferase	*Ae. aegypti*	Male/female	ND[b]	Embryo	Shotkoski *et al*. (1996)
Ferritin LCH	Luciferase	*Ae. aegypti*	ND[b]	ND[b]	Cell culture	Pham and Chavez (2005)
CecB	Luciferase	*Ae. aegypti*	Female	Fat body	Lipopolysaccharide inducible	Isoe *et al*. (2007)
nanos	*Mos1* transposase	*Ae. aegypti*	Female	Ovaries/embryos	Embryo	Adelman *et al*. (2007)
AgApy	*LacZ*	*An. stephensi*	Female/male	Salivary glands	Constitutive	Lombardo *et al*. (2005)

[a]These columns indicate the species in which the DNA was tested and summary information on the expression profile.
[b]Not determined.

accumulated ectopically in the lobes of female salivary glands. Abundant expression in mosquito salivary glands was achieved only recently using a prompter from the anopheline antiplatelet protein (AAPP), a 30 K protein family gene in *An. stephensi* (Yoshida and Watanabe, 2006). The amount of transgene product (DsRed protein) was calculated to be 25 ng per pair of salivary glands. This is ~5% of the total salivary gland protein level, and a >1000-fold higher level of expression than reported for other salivary gland-specific gene promoter–reporter constructs (Coates *et al.*, 1999; Lombardo *et al.*, 2005). Unlike the endogenous AAPP mRNAs, whose abundance peaks at 48 h postblood meal (hPBM), accumulation of DsRed mRNA reached a high level as early as 24 hPBM and maintained this expression level for the subsequent 24 h. One explanation of this abundance of the DsRed mRNA may be due to a high stability of its mRNA.

Genes whose products are involved in digestion of the blood meal are induced within a short time after feeding. A number of trypsins, peptidases, other enzymes, and transport proteins are recognized as being involved in this developmental phase. There are early and late phases of digestion that vary among the different mosquito species in the specific timing and genes involved (Barillas-Mury *et al.*, 1995; Müller *et al.*, 1993). The best characterized genes are those that encode carboxypeptidases (CPs, Table 2.3). The endogenous genes are upregulated within a few hours after a blood meal and likely are involved in the primary digestion of the blood meal (Edwards *et al.*, 1997, 2000). Functional promoters have been characterized from genes derived from *Ae. aegypti* and *An. gambiae* (*AeCP* and *AgCP*, respectively; Kim *et al.*, 2004; Moreira *et al.*, 2000). While immunoblot data indicate that carboxypeptidase promoter-driven transgenes could result in the accumulation of ~2 ng of reporter gene protein per gut at 24 hPBM (Moreira *et al.*, 2000), the accumulation of heterologous gene products may be so low that it cannot be detected, and proper function of the promoter was established by gene amplification of mRNAs (Kim *et al.*, 2004).

A set of studies investigated the ability of CP genes to function in heterologous hosts. *Aedes aegypti* and *An. gambiae* bear only a distant evolutionary relationship and may have last shared a common ancestor 140–180 million years ago (Krzywinski *et al.*, 2006). Despite an apparent lack of sequence similarity between the *AeCP* and *AgCP* putative promoters, both can drive robust expression of a luciferase reporter mRNA and protein in a blood-inducible manner in *Ae. aegypti* (Moreira *et al.*, 2000). Interestingly, the native induction in *An. gambiae* of *AgCP* by a blood meal is rapid, 3 hPBM, but is much slower, 24 hPBM, when the sequences are integrated in *Ae. aegypti*. The *AeCP* promoter functioned in *An. gambiae* to drive reporter gene expression in the posterior midgut beginning ~24 hPBM (Kim *et al.*, 2004).

The fat body tissues of mosquitoes often express genes whose products are localized or transported through the hemolymph. The regulatory region of a vitellogenin-encoding gene (*Vg1*) of *Ae. aegypti* directed high levels of expression resulting in the abundant accumulation of the reporter gene product in the hemolymph of blood-fed female mosquitoes (Kokoza *et al.*, 2000). The reporter gene used in these studies encoded a defensin, and the protein persisted in the hemolymph of blood-fed female mosquitoes for 20–22 days after a single blood feeding. Control sequences of *AsVg1* directed blood meal stimulated sex- and tissue-specific expression of a reporter gene in transgenic adult female *An. stephensi* (Nirmala *et al.*, 2006). DNA fragments encompassing the 850 bp immediately adjacent to the 50-end of the gene and the 30-end untranslated region are sufficient to direct this expression. The control sequences of an *An. gambiae* vitellogenin-encoding gene, *VgT2*, directed correct expression of a reporter gene in a tissue-, stage-, and sex-specific manner in *An. stephensi* (Chen *et al.*, 2007). Furthermore, multiple blood meals resulted in persistent expression of the reporter gene, making the promoter a good candidate for directing the abundant accumulation of exogenous gene products.

Detailed analyses of transcription factor-mediated gene expression in mosquitoes have been carried out in *Ae. aegypti* by Raikhel and colleagues (Attardo *et al.*, 2003; Cho *et al.*, 2006; Park *et al.*, 2006). Following blood feeding, genes involved in the reproductive cycle, including those encoding yolk protein precursors such as vitellogenin and the vitellogenin receptor (*VgR*) are induced. *VgR* is involved in the receptor-mediated endocytosis of vitellogenin. A 1.5 kb fragment comprising 5′-end, putative *cis*-active *VgR* DNA was sufficient for correct female and ovary-specific expression of a transgene. The fragment includes binding sites for the products of the gene, E74 and BR-C, involved in the ecdysone response, as well as other sites for transcription factors determining correct tissue- and stage-specific expression. The 5′-end DNA of the vitellogenin-encoding gene has multiple sites for GATA-binding factors that are necessary for abundant expression. Remarkably, the products of a GATA factor-binding gene (AaGATAr) function as repressors prior to blood meal induction of vitellogenin-encoding genes. Of great interest was the demonstration that amino acid signaling through the nutrient-sensitive target of rapamycin (TOR) pathway is essential for the activation of YPP gene expression (Hansen *et al.*, 2004).

A number of other genes have been assayed functionally by transgenesis in mosquitoes. Much of the original work developing transformation vectors relied on promoters derived from *D. melanogaster*. Helper plasmids have the transposase open reading frames under control of heat shock promoters (HSP70 and HSP87) and a number of actin gene promoters have been used to drive the expression of marker genes (Catteruccia *et al.*, 2003; Pinkerton *et al.*, 2000). An EGFP marker gene under the control of the *act88F* gene promoter of *D. melanogaster* inserted into the genome of *C. quinquefasciatus* showed expression restricted to the flight muscle (Allen and Christensen, 2004). *β2*-tubulin genes encode a protein that is expressed specifically in the male gonads. The promoter of the *An. gambiae* orthologous gene was used to drive EGFP expression in *An. stephensi* (Catteruccia *et al.*, 2005). Fluorescence was observed principally in the male gonads of the transgenic lines recapitulating the expression profile of the endogenous gene. Similar experiments were done with the $\beta 2$ orthologous gene of *Ae. aegypti* (Smith *et al.*, 2007). As with the previous study, expression of a fluorescent reporter gene marked sperm that were detected in spermathecae of inseminated females. The *nanos* (*nos*) gene is expressed in females and is localized in the oocytes and is responsible for determining the anterior–posterior axis in developing embryos. The *nos* orthologous gene of *Ae. aegypti* was able to control sex- and tissue-specific expression of exogenously derived *MosI* transposase-encoding DNA (Adelman *et al.*, 2007). Transgenic mosquitoes expressed transposase mRNA in abundance and exclusively in the female germ cells. In addition, transgene mRNA was deposited in developing oocytes and localized and maintained at the posterior pole during early embryonic development.

Conditional and targeted expression of transgenesis in specific tissues can be achieved by using a tetracycline-regulated system. This system of conditional expression is based on transcriptional activators (TA) or reverse transactivator (rtTA) (Gossen and Bujard, 1992; Urlinger *et al.*, 2000) that respond to the antibiotic, tetracycline (tet), or an analog, doxycycline (dox). While TA binds to tet operator-derived response elements (TetO) in the absence of tet and permits transcription, rtTA binds in the presence of tet. This conditional expression system was tested in *An. stephensi* using *An. gambiae* SRPN10 promote (Lycett *et al.*, 2004). Two transgenic lines were generated, one expressing TA or rtTA controlled by a tissue-specific *SRPN10* promoter and another, expressing the *LacZ* gene under the control of TetO. Progeny obtained from a cross between the two transgenic lines show inducible *LacZ* expression in pericardial cells, hemocytes and epithelial midgut cells regulated by the SRPN10 promoter as well as by the presence or absence of dox. Surprisingly, a blood meal-specific increase in the percentage of *LacZ* expressing hemocytes was seen. Gene regulation by rtTA was stronger than TA in this study.

2. Expression of antipathogen effector genes

A major application of characterized *cis*-acting control DNA is to regulate the expression of transgenes that target specific mosquito-borne pathogens. The concept is that it should be possible to use transgenesis technologies to make mosquitoes resistant to a specific parasite or virus, and then use that mosquito in disease intervention strategies to control transmission (Meredith and James, 1990). Key to the successful development of these effector molecules is the utilization of *cis*-acting regulatory sequences that direct their expression to the pathogens at the right time and place, and in sufficient abundance to have an effect. The major pathogens that are targeted are the dengue viruses and malaria parasites. Dengue viruses have a positive-strand RNA genome, and a series of elegant studies have shown that these viruses are vulnerable to RNAi-mediated destruction (Sanchez-Vargas *et al.*, 2004). This work culminated in the production of transgenic *Ae. aegypti* that use the *AaCP* gene promoter to drive an RNAi-inducing transgene to reduce significantly the level of virus in mosquitoes (Franz *et al.*, 2006).

A wider variety of antiparasite effector molecules is available to attack malaria parasites (Nirmala and James, 2003). These include those that interfere with parasite ligands or mosquito-encoded molecules that facilitate tissue recognition by the parasite (pathogen "receptors"). Others interfere with parasite gene expression or elicit elevated insect immune responses. Finally, toxins also have been expressed that can differentially target the parasites. The most successful antiparasite effector genes used the CP-encoding promoters to target midgut-stage parasites. CP promoters activated <3 hPBM in the midgut can be used to interfere with parasite gametes, zygote and ookinetes. A 12 amino acid peptide, designated SM1, for salivary gland and midgut binding, was used to develop a transgene consisting of four SM1 units driven by the *AgCP* promoter (Ito *et al.*, 2002). SM1 expressed in transgenic females following an infectious blood meal bound to the luminal surface of the midgut, inhibiting significantly parasite–epithelium interactions and midgut invasion. Transgenic *An. stephensi* mosquitoes expressing a bee venom phospholipase A2 gene (PLA2) with *AgCP* as the promoter reduced parasite oocyst formation and greatly impaired transmission of the parasite to mice (Moreira *et al.*, 2002). Follow-up studies with the regulatory DNA of the *An. gambiae* adult peritrophic matrix protein 1 (AgAper1) promoter fused to PLA2 led to the accumulation of PLA2 in midgut epithelial cells before a blood meal and its release into the lumen upon blood ingestion greatly affecting oocyst formation (Abraham *et al.*, 2005). Transient and stable transformation studies using single-chain antibodies (scFv) targeting important parasite ligands or other expression products have been shown to reduce intensities of infection of specific developmental stages (de Lara Capurro *et al.*, 2000; Yoshida *et al.*, 1999). Stable transgenics used mosquito vitellogenin-encoding (*Vg1*) and *D. melanogaster ubiquitin* gene promoters in parasite-infected *Ae. aegypti* (Jasinskiene *et al.*, 2007). These reagents have the benefit of targeting the parasite directly and thus are expected to not have a major negative effect on the fitness of the mosquito expressing them.

Manipulating the innate immune system of mosquitoes may be a possible approach to affect negatively their capacity to serve as parasite hosts. Immune system-based effector genes encoding antimicrobial peptides are effective against *Plasmodium in vitro* (Christophides *et al.*, 2002). Transgenes consisting of one of two representatives of the major peptide families comprising defensins and cecropins doubled the resistance exhibited by *Ae. aegypti* to the Gram-negative bacterium, *Enterobacter cloacae* (Shin *et al.*, 2003). Furthermore, transgenic *Ae. aegypti* strains overexpressing Defensin A inhibited oocyst growth of *Plasmodium gallinaceum*, and two independent lines of transgenic *An. gambiae* containing the *An. gambiae* cecA cDNA driven by the *AeCP* promoter showed a 60% reduction in the number of oocysts (Kim *et al.*, 2004; Shin *et al.*, 2003).

Maintaining large numbers of colonies, issues with pair matings and the need for blood meals at every generation make difficult standard genetic analyses with mosquitoes. Reverse genetic approaches, such as RNAi-mediated expression ablation, provide a way to circumvent these difficulties (Shin *et al.*, 2003). The assessment of gene function based on global expression patterns is often validated by a more direct method such as RNAi. Gene silencing can be achieved by injecting double-stranded RNA (dsRNA) into adult mosquito hemolymph (Blandin *et al.*, 2002) or by injecting recombinant Sindbis virus expressing specific dsRNAs (Attardo *et al.*, 2003). In the first study, dsRNA targeting transcripts of the antimicrobial peptide, defensin, showed the significance of this gene product in defense against infections with Gram-positive bacteria but not *P. berghei* infection, in contrast to the studies with *P. gallinaceum* infections of *Ae. aegypti* (Lowenberger *et al.*, 1999; Shin *et al.*, 2003). In the second experiment, knockdown of the AaGATAr gene revealed its role in regulating *Vg* gene expression and responses to 20-hydroxyecdysone.

Gene products can be knocked down efficiently in the mosquito fat body tissues and hemocytes. However, higher amounts of dsRNA are required to achieve knockdown in the salivary glands, presumably due to the poor permeability of the glands to nucleic acids (Boisson *et al.*, 2006). Injection of microgram quantities of dsRNA into *An. gambiae* showed that ablation of the products of *AgApy*, a gene encoding a platelet antiaggregating factor, apyrase, lengthened the duration of the probing behavior.

Genes that play important roles in vector-parasite interactions also are being identified using RNAi methods. The antiparasitic role of a complement-like protein, TEP1, was first demonstrated in *An. gambiae* by RNAi-mediated knockdown (Blandin *et al.*, 2004). Ablation of TEP1 transcription products prevents a type of mosquito parasite refractoriness based on melanotic encapsulation of the pathogens. Another study showed that two mosquito genes had complementary effects, with WASP having a negative and ApoII/I a positive role in parasite development (Mendes *et al.*, 2008). Resistance to malaria parasite infection in a nonvector species *An. quadriannulatus* was shown to result from an innate immune response in which three genes, LRIM1, LRIM2, and TEP1, play a major role (Habtewold *et al.*, 2008). Knockdown of these genes rendered the mosquito permissive to parasite infection. RNAi-mediated inhibition of the expression of the chitin synthase gene was used to probe the function of the peritrophic matrix (PM) in affecting pathogen invasion (Kato *et al.*, 2008). While PM loss did not have an impact on the development of the filarial worm, *Brugia pahangi*, or the dissemination of dengue virus, *P. gallinaceum* oocyst numbers were reduced. Interestingly, the absence of a PM had no effect on reproductive fitness. These studies emphasize the utility of RNAi approaches for dissecting complex mosquito phenotypes associated with development and vector competence.

Expression of transgenes in mosquitoes must not impose too great a genetic load if they are to be useful in control strategies. Two aspects of fitness could affect the success of a genetic control program based on transgenic mosquitoes, and for discussion purposes, they can be referred to as the "optimal" and "suboptimal" fitness of the transgenic strains (Scott, 2006). Optimal refers to those features of fitness that maximize the reproductive rate of the strain and are evident under laboratory rearing conditions in the absence of competition. Suboptimal conditions are present in the wild and include competition. All fitness estimates are relative and optimal reproductive success of the transgene-bearing lines must be referenced to the nontransgenic strain from which they were derived. In the simplest terms, female mosquitoes must be physically able to locate and feed on hosts, and recognize and use appropriate oviposition sites. Male mosquitoes must be able to locate and mate with females. Both sexes must be able to complete development without significant lethality in the adult and subadult stages. In addition to inbreeding depression, two other mechanisms associated directly with transgenesis, insertional mutagenesis and metabolic load caused by the expression of the transgene, may affect mosquito fitness. Insertional mutagenesis may cause a fitness reduction if integration of the transgene into the host genome results in the partial or complete disruption of an endogenous gene function at the insertion site. Expression of the transgene may be detrimental to mosquitoes if the gene product is toxic or gene translation usurps resources needed for normal reproductive functions. No studies have examined yet the effect of insertional mutagenesis and expression of the transgenes on mosquito fitness.

Fitness measurements of transgenic mosquitoes stimulate considerable debate (Marrelli *et al.*, 2006), and experiments performed to measure loads imposed by the insertion of exogenous genes yield data supporting contradictory conclusions. Catteruccia *et al.* (2003) found that *An. stephensi* transformed with a *Minos*-based transposon construct with the *D. melanogaster Actin5C* or *hsp70* gene promoters driving the expression of EGFP exhibited a large fitness load in three of four transgenic lines in comparison to nontransgenic animals. Furthermore, *Ae. aegypti* transformed with a *Hermes* transposon carrying the *Actin5C* promoter linked to EGFP also exhibited significant fitness reductions (Irvin *et al.*, 2004). In contrast, Moreira *et al.* (2004) found that *An. stephensi* transformed with a *piggyBac*-based transposon, EGFP reporter, and the *AgCP* promoter driving expression of the SM1 peptide effector gene had no detectable fitness load. The discrepancy in the results of these three studies was attributed to inbreeding depression (making homozygous one or more recessive alleles with negative fitness phenotypes) in that the Moreira study used hemizygous (one copy of the transgene, no alternative allele) transgenic animals to measure fitness, whereas the other two used homozygous individuals (Marrelli *et al.*, 2006). The nature of the expressed protein itself is a crucial factor for fitness. For example, although no effect on fitness was observed for mosquitoes expressing SM1, mosquitoes expressing PLA2 were clearly less fit and less fertile than wild-type (Abraham *et al.*, 2005). A recent report argues that there may be a tradeoff between fitness effects of the transgene and parasite infection (Marrelli *et al.*, 2007). Mosquitoes with transgenes expressing antiparasite molecules may outperform wild-type mosquitoes when both are infected with parasites.

C. Genomics

1. Comparative genomics

Comparative genomic analyses exploit whole genome sequences to identify novel and common genes among species, increase our knowledge of insect evolution, and complement the work done with single or small numbers of genes. In addition to gene discovery, analyses of common genes also can provide insights into mosquito gene regulation. Comparisons were done first with *An. gambiae* and *D. melanogaster*, which are estimated to have last shared a common ancestor some 250 million years ago (Holt *et al.*, 2002; Zdobnov *et al.*, 2002). While nearly half of the annotated genes appear to have orthologs in both species, sequence identity among these was ~56%. The authors point out that this number is lower than the number seen between humans and a fish that diverged ~450 million years ago. This was interpreted to indicate that these insects diverged at a higher rate than vertebrates.

The addition of the *Ae. aegypti* (Nene *et al.*, 2007) and *C. quinquefasciatus* (Vectorbase) genomes provides an opportunity to perform comparative genome analyses between mosquito species that diverged 140–180 million years ago (Krzywinski *et al.*, 2006). The genome size of *Ae. aegypti* is five times larger than *An. gambiae*, most likely as a result of a large number of encoded transposable elements and a higher content of repetitive DNA (Waterhouse *et al.*, 2008). Remarkable species-specific expansions of gene families encoding odorant-binding proteins, cytochrome P450s, secreted salivary gland proteins and genes whose products have domains characteristic of cuticle illustrate major differences among these organisms.

2. Functional genomics

Global gene expression analyses in mosquitoes are particularly useful in comparing expression profiles of nonblood-fed females before and after a blood meal, females and males, infected and noninfected mosquitoes and insecticide-resistant and susceptible strains. Genome-wide changes in expression levels were studied in the midgut of *Ae. aegypti* following a blood meal using cDNA microarrays comprising clones from expressed sequence tags (ESTs) (Sanders *et al.*, 2003). Not too surprisingly, significant regulation was seen in classes of genes involved in nutrient absorption and metabolism, cellular stress responses, ion balance, and formation of the peritrophic matrix. A larger-scale project with *An. gambiae* showed that as many as 33% of all annotated genes vary the abundance of their corresponding transcription products within 24h after a blood meal (Marinotti *et al.*, 2005). Major changes in transcript abundance were seen in genes encoding proteins involved in digestion, oogenesis, and locomotion. A comparison of gene expression profiles of *An. gambiae* adults at discrete times after a blood meal, in senescent adults and males was made (Marinotti *et al.*, 2006). Approximately 22% of the genes had sex-dependent regulation. Females devote the majority of their metabolism to blood digestion and egg formation within 3 hPBM and downregulate gene involved in flight and response to environment stimuli. The majority of changes in expression are evident over the first three days after a blood meal, when as many as 50% of all genes showed significant variation in transcript accumulation. After laying the eggs (between 72 and 96 hPBM), mosquitoes return to a nongonotrophic stage, similar to that of 3-day-old nonblood-fed females. Data from these studies was compiled in a database for its use to study gene expression in a particular stage of the mosquito (Dissanayake *et al.*, 2006; Vectorbase). Additionally, algorithms to identify conserved *cis* regulatory motifs in genes coordinately expressed were included in the database. A proteomics approach using two-dimensional gel electrophoresis also was used to look at sex-specific and blood meal-induced proteins in *An. gambiae* midguts (Prévot *et al.*, 2003). Of ~375 observed proteins, samples from males showed ten not evident in sugar- or blood-fed females. Female midguts contained 23 proteins not found in males, eight of which were specific to sugar-fed and ten to blood-fed females. These data need to be followed-up with mass spectroscopy analyses to identify the specific proteins.

A detailed analysis of transcript abundance between the male and female *An. gambiae* midguts was carried out as well as for four regions in the female midgut (Warr *et al.*, 2007). Significant differences were found between the sexes in expression profiles and levels in genes involved in digestion and immunity. Furthermore, each anatomical region of the female midgut has a characteristic expression profile, for example, the posterior midgut expresses genes encoding digestive enzymes, while the anterior midgut expresses in abundance antimicrobial peptides and other immune gene products. The functional diversity of the female midgut observed was attributed to the requirements for dealing with blood meals and possible microbial infections that this feeding behavior would promote. Regional specification of the larval midgut also was demonstrated by microarray analyses (Ovieda *et al.*, 2008). The data support the conclusions that protein and carbohydrate metabolism and absorption take place in the gastric caeca and posterior midgut, while lipids are processed in the anterior midgut. Similar to the adult female midgut, transcripts corresponding to antimicrobial peptides and enzymes involved in detoxifying xenobiotics were localized to the gastric caeca and anterior midgut.

Bioinformatic and data mining searches were applied to the *An. gambiae* genome as soon as it was available. One of the more productive analysis identified genes encoding odorant-binding proteins (OBPs) and odorant receptors (ORs) (Biessmann *et al.*, 2002; Fox *et al.*, 2001). It is anticipated that this work could lead to the development of novel ways of preventing mosquitoes from finding and feeding on humans. Arrestins and $G\alpha$-encoding genes potentially involved in olfactory signal transduction also were identified. One of these ORs, AgOr7, is highly conserved in insects, and specifically in *An. gambiae*, and *Ae. aegypti* (Melo *et al.*, 2004; Pitts *et al.*, 2004). Sequence conservation and similar expression characteristics indicate an important and common olfactory function of these mosquito receptors. Expression in the proboscis of the orthologous genes in both mosquito species differs from the expression profile of the *D. melanogaster* gene, DOr83b, likely indicating adaptations to sensing different food substrates. Twenty-four OR-like gene products are located on the proboscis, which is consistent with a role as an accessory olfactory organ (Kwon *et al.*, 2006). The *An. gambiae Arrestin1* (*AgArr1*) rescued an olfactory deficit due to mutations of orthologous gene, *DmArr1*, in *D. melanogaster*, but the mutation-linked larval behavioral deficit was not rescued (Walker *et al.*, 2008). Additionally both *AgArr2* and *DmArr2* were unable to rescue the *DmArr1* mutation in fruit flies, suggesting a nonredundant function of the two different arrestins. In a different study, *An. gambiae* larval olfactory behaviors were examined using whole animal activity assays (Xia *et al.*, 2008). Strong responses in larvae were associated with compounds related to cresol that are produced from the decay of organic matter, which constitute the food of the larvae. These responses could be attenuated by ablation of the larval antennae. Four ORs were found that are specific to the larvae. Functional analysis of nine larval AgORs was performed with 82 odorants using *Xenopus* oocyte responses (Wetzel *et al.*, 2001). Multiple ORs responding to a single odorant indicate a combinatorial coding mechanism to encode odorant information in the larvae. Interestingly, both behavioral and oocyte responses of larval-specific AgOR40 to adult insect repellent, DEET indicate either the presence of additional DEET-sensitive receptors involved in adults or an indirect role of DEET as a behavioral repellent. Recently, a study revealed that insect ORs comprise a new class of ligand-activated nonselective cation channels (Sato *et al.*, 2006). ORs from *D. melanogaster*, *Bombyx mori*, and *An. gambiae* were expressed with their respective coreceptors in heterologous system where upon activation, they show increase in intracellular $Ca2+$. This activation does not involve G-protein signaling, providing evidence for a unique olfactory mechanism acquired by insects. Microarray-based studies of OBPs identified a number that are expressed at higher levels in either female or male antennae and palps (Biessmann *et al.*, 2005). Changes in OBP expression following a blood meal may reflect changes in the behavioral activity of females as the progress from blood feeding to nectar feeding.

Functional genomics also have been applied to define mosquito genes involved in interactions with malaria parasites. DNA microarrays were used to evaluate midgut responses to *P. berghei* and *P. falciparum* infections in *An. gambiae* (Vlachou *et al.*, 2005). Remarkably, >7% of the assessed mosquito transcriptome is regulated differentially following infection. Ookinete penetration of the midgut epithelium stimulates genes involved in actin- and microtubule/cytoskeleton-mediated and extracellular-matrix remodeling. Not surprisingly, other induced genes encode products involved in innate immunity and apoptosis. Coupled analyses with RNAi-mediated gene silencing identified antagonists and agonists of actin formation and provide evidence to support the conclusion that actin polymerization inhibits parasite invasion. Furthermore, the immune responses of the mosquito to the two parasites are quite diverse, and while some antimicrobial factors are induced in common upon infection, parasite-specific factors also are seen.

Infection by arthropod-borne viruses also stimulates changes in gene expression in mosquitoes. Although culicine mosquitoes (*Aedes*, *Culex*) transmit a large number of different flaviviruses (dengue, yellow fever) and bunyaviruses (Chikungunya, Rift Valley Fever), anopheline mosquitoes are rarely associated with virus transmission. A notable exception is o'nyong-nyong, and *An. gambiae* infected with this virus were assayed with microarrays carrying ~20,000 cDNAs (Sim *et al.*, 2005). Comparisons with uninfected mosquitoes revealed 18 genes whose expression levels varied significantly up or down. Although additional studies have to be done to identify the precise role of these genes in response to infection, the authors speculate that some of them may reflect efforts of the mosquitoes to resist infection or the action of the viruses to harness the biosynthetic machinery of the cells. Microarray analysis of *Ae. aegypti* infected with dengue viruses provides evidence that mosquitoes do respond to viral infection by inducing expression of genes involved in immune response, specifically those associated with the *Toll* pathway (Xi *et al.*, 2008). These authors also noted that the natural intestinal fauna of the mosquito may be a factor in attenuating dengue viral infection by providing a constitutive level of expression of genes in the *Toll* immune pathway.

Christensen and colleagues looked at expression patterns in mosquitoes infected with filarial worms (Aliota *et al.*, 2007; Bartholomay *et al.*, 2004). Many EST sequences from immune response-activated hemocyte libraries of *Ae. aegypti* and *Armigeres subalbatus* corresponded to immunity-related genes, some of which had strong similarity to genes involved in vertebrate innate immunity. In a follow-up study, the transcriptional response of *Ar. subalbatus* to *B. malayi* was shown to be complex and tissue specific.

Insecticide resistance in mosquitoes is a major factor driving efforts to develop genetics-based control strategies. However, insecticides still are the best tools available today to fight epidemic mosquito-transmitted disease outbreaks, and therefore there are compelling reasons to understand the molecular bases of resistance. This knowledge aids in particular the development of diagnostic tools to monitor resistance as part of programs that seek to manage resistance and therefore prolong the efficacy of specific chemicals. A cDNA microarray analysis of insecticide-resistant and susceptible *An. gambiae* detected 77 differentially transcribed ESTs (Vontas *et al.*, 2005). These included representatives of genes in expected families, for example, cytochrome P450s, a carboxylesterase, UDP-glucuronosyltransferases and nitrilases, and a number, including those encoding peptidases, ion exchangers and enzymes involved in lipid and carbohydrate metabolism, that were not anticipated to be linked to insecticide resistance. Furthermore, the array developed for *An. gambiae* could be used with reduced hybridization stringency to analyze resistance in *An. stephensi* (Vontas *et al.*, 2007). Although background was high, ~7000 significant signals were detected, 36 genes differed in expression levels among resistant and susceptible strains, including glutathione *S*-transferases, esterases, cytochrome P450s, and peroxidases.

One of the advantages of working with *D. melanogaster* is the wealth of information available on gene expression profiles throughout the development of the animal (http://flybase.bio.indiana.edu/). This type of information is now being put together for some mosquitoes, although much work still needs to be done. Spatial and temporal gene expression analysis of the *An. gambiae* life cycle reported genes involved with specific developmental stages (Koutsos *et al.*, 2007). Not too surprisingly, expression of genes known or annotated to be related functionally correlates with specific developmental stages or tissues. Furthermore, although a positive correlation was found for expression profiles among orthologous genes in the mosquito and fruit fly, the profiles did not correlate with coding sequence similarity. These data support the hypothesis that selection pressures allow independent evolution of expression properties and coding sequences.

Few detailed analyses have been made of individual mosquito genes involved in early development. While the expression profiles of the *nanos* and *oskar* (*osk*) genes are conserved generally, there are some differences noted in temporal and spatial distribution (Adelman *et al.*, 2007; Goltsev *et al.*, 2004; Juhn and James, 2006). Specifically, while mosquito *oskar* mRNAs were found to localize to the posterior end in early-stage embryos, *An. gambiae* also has transient localization in the anterior.

The development of rapid and inexpensive sequencing techniques now allows comprehensive analyses of transcriptomes of whole animals, development stages and isolated tissues. Millions of short cDNAs can be sequenced garnering information on transcript complexity and abundance. Perhaps no better example exists of the power of this approach than the efforts of Ribeiro and colleagues to define the "sialomes," complete expression profiles of the salivary glands in mosquitoes and other hematophagous insects. Translation of most of the transcripts identified was confirmed by SDS-PAGE and Edman degradation. Female mosquitoes secrete a variety of proteins such as vasodilators, anticoagulants, and platelet anti-aggregating factors while taking a blood meal. Full knowledge of the transcriptome enables discovery of genes involved in various physiological processes associated with blood feeding. The data also provides insight into the evolution of the salivary gland protein families. The new genes discovered are too numerous to recount here, so the readers are referred to a number of key papers and a recent review (Calvo *et al.*, 2007; Francischetti *et al.*, 2002; Ribeiro, 2003; Ribeiro *et al.*, 2004, 2007; Valenzuela *et al.*, 2002). Key conclusions are that genes that encode proteins in phylogenetically widespread families are adapted for major roles in salivary gland function. Furthermore, there are salivary gland products derived from protein families present only in blood-feeding Nematocera. Within this group, there are families of proteins found only in the culicines or anophelines. Particular note is made of finding evidence of protein families of prokaryotic origin, most likely due to horizontal transfer.

Transcriptome studies of mosquito responses to parasite infection complement those done with whole-genome or selected microarrays. Sequencing of 1485 random clones obtained from subtracted cDNA libraries of *An. stephensi* infected with *P. berghei* identified >1100 unique reads (Srinivasan *et al.*, 2004). These provided information on both mosquito and parasite genes that were regulated during infection. For example, a mosquito gene encoding a fibrinogen domain was induced coincidently with the parasite transition from ookinete to oocyst. Using the same approach, Ancaspase-7, an *Anopheles* effector caspase activated during *Plasmodium* invasion of midgut was discovered (Abraham *et al.*, 2004). A similar approach was used to identify genes in *Ae. aegypti* midguts expressed differentially in mosquitoes susceptible or refractory to the avian malaria parasite *P. gallinaceum* (Chen *et al.*, 2004). Approximately 2.3% of 1200 midgut cDNA were expressed differentially between the susceptible and refractory mosquito populations. Of particular note, four genes corresponded to marker genes used in quantitative-trait analysis of parasite refractoriness in the same mosquito (Severson *et al.*, 2002).

III. CONCLUDING REMARKS

Recent years have witnessed rapid progress in the development of new techniques to study gene expression in mosquitoes. Development of transgenesis and a growing complement of effector molecules bring closer the promise of the use of a genetic strategy to control the transmission of mosquito-borne diseases. The expression of fluorescent proteins facilitates the screening of transformants and functional analysis of *cis*-acting regulatory regions of genes. However, the expression levels of heterologous genes in transgenic mosquitoes are usually low, which might affect their utility in genes designed to block pathogen transmission. Identification and characterization of more robust gene promoters are required to overcome these low expression levels. Also, more tissue-specific gene promoters are required to express multiple effector molecules in the same transgenic mosquito, which is expected to increase the effect of antipathogen phenotype as well as decrease the possibility of development of pathogen resistance to the effectors. There is yet no complete inhibition of pathogen transmission in transgenic mosquitoes.

Rapid progress has been seen in the field of genomics due to the availability of three mosquito genomes. Genes differentially expressed following blood meals, parasite infection and as a consequence of insecticide resistance have been studied extensively. Remarkable expansions of major gene families involved in odor perception, immune responses, and salivary gland products provide materials for years of future research. Both the genomics and transcriptome studies have been complemented immensely by studies using RNAi for gene silencing. RNAi allows characterization of genes *in vivo* which can later be targeted for transmission blocking studies. It also helps in understanding the mechanism of gene regulation.

The field of mosquito molecular biology would benefit from the sequencing of additional mosquito genomes. Population-level studies of gene variation also could provide insights into aspects of behavior that make these insect such dangerous animals. Practically, better transgenesis tools are needed if the techniques are ever to see the widespread applications that are characteristic of research with *D. melanogaster*. Methods for long-term storage of strains also are needed. Finally, the field would be strengthened greatly by the recruitment of new, young and creative researchers looking for topics in translational science.

Acknowledgments

Original work by the authors is supported by grants from the Natural Science Foundation of China (No. 30771871 to X-GC), the National Institute of Allergy and Infectious Diseases, National Institutes of Health (AAJ), and in part by a grant to the Regents of the University of California from the Foundation for the National Institutes of Health through the Grand Challenges in Global Health initiative (GM and AAJ). Lynn M. Olson assisted in the preparation of the manuscript.

References

Abraham, E. G., Islam, S., Srinivasan, P., Ghosh, A. K., Valenzuela, J. G., Ribeiro, J. M., Kafatos, F. C., Dimopoulos, G., and Jacobs-Lorena, M. (2004). Analysis of the Plasmodium and Anopheles transcriptional repertoire during ookinete development and midgut invasion. *J. Biol. Chem.* **279,** 5573–5580.

Abraham, E. G., Donnelly-Doman, M., Fujioka, H., Ghosh, A., Moreira, L., and Jacobs-Lorena, M. (2005). Driving midgut-specific expression and secretion of a foreign protein in transgenic mosquitoes with AgAper1 regulatory elements. *Insect Mol. Biol.* **14,** 271–279.

Adelman, Z. N., Jasinskiene, N., and James, A. A. (2002). Development and applications of transgenesis in the yellow fever mosquito, *Aedes aegypti*. *Mol. Biochem. Parasitol.* **121,** 1–10.

Adelman, Z. N., Jasinskiene, J., Vally, K. J. M., Peek, C., Travanty, E. A., Olson, K. E., Brown, S. E., Stephens, J. L., Knudson, D. L., Coates, C. J., and James, A. A. (2004). Formation and loss of large, unstable tandem arrays of the *PiggyBac* transposable element in the yellow fever mosquito, *Aedes aegypti*. *Transgenic Res.* **13,** 411–425.

Adelman, Z. N., Jasinskiene, N., Onal, S., Juhn, J., Ashikyan, A., Salampessy, M., MacCauley, T., and James, A. A. (2007). *nanos* gene control DNA mediates developmentally-regulated transposition in the yellow fever mosquito, *Aedes aegypti*. *Proc. Natl Acad. Sci. USA* **104,** 9970–9975.

Aliota, M. T., Fuchs, J. F., Mayhew, G. F., Chen, C. C., and Christensen, B. M. (2007). Mosquito transcriptome changes and filarial worm resistance in *Armigeres subalbatus*. *BMC Genomics* **18,** 463.

Allen, M. L., and Christensen, B. M. (2004). Flight muscle-specific expression of act88F: GFP in transgenic *Culex quinquefasciatus* Say (Diptera: Culicidae). *Parasitol. Int.* **53,** 307–314.

Allen, M. L., O'Brochta, D. A., Atkinson, P. W., and Levesque, C. S. (2001). Stable, germ-line transformation of *Culex quinquefasciatus* (Diptera: Culicidae). *J. Med. Entomol* **38,** 701–710.

Atkinson, P. W., and James, A. A. (2002). Germ-line transformants spreading out to many insect species. *Adv. Genet.* **47,** 49–86.

Attardo, G. M., Higgs, S., Klingler, K. A., Vanlandingham, D. L., and Raikhel, A. S. (2003). RNA interference-mediated knockdown of a GATA factor reveals a link to anautogeny in the mosquito *Aedes aegypti*. *Proc. Natl Acad. Sci. USA* **100,** 13374–13379.

Attardo, G. M., Hansen, I. A., and Raikhel, A. S. (2005). Nutritional regulation of vitellogenesis in mosquitoes: Implications for anautogeny. *Insect Biochem. Mol. Biol.* **35,** 661–675.

Barillas-Mury, C. V., Noriega, F. G., and Wells, M. A. (1995). Early trypsin activity is part of the signal transduction system that activates transcription of the late trypsin gene in the midgut of the mosquito, *Aedes aegypti*. *Insect Biochem. Mol. Biol.* **25,** 241–246.

Bartholomay, L. C., Cho, W. L., Rocheleau, T. A., Boyle, J. P., Beck, E. T., Fuchs, J. F., Liss, P., Rusch, M., Butler, K. M., Wu, R. C., Lin, S. P., Kuo, H. Y., *et al.* (2004). Description of the transcriptomes of immune response-activated hemocytes from the mosquito vectors *Aedes aegypti* and *Armigeres subalbatus*. *Infect. Immun.* **72,** 4114–4126.

Berg, D. E., Schmandt, M. A., and Lowe, J. B. (1983). Specificity of transposon Tn5 insertion. *Genetics* **105,** 813–828.

Berghammer, A. J., Klingler, M., and Wimmer, E. A. (1999). A universal marker for transgenic insects. *Nature* **402,** 370–371.

Biessmann, H., Walter, M. F., Dimitratos, S., and Woods, D. (2002). Isolation of cDNA clones encoding putative odourant binding proteins from the antennae of the malaria-transmitting mosquito, *Anopheles gambiae*. *Insect Mol. Biol.* **11,** 123–132.

Biessmann, H., Nguyen, Q. K., Le, D., and Walter, M. F. (2005). Microarray-based survey of a subset of putative olfactory genes in the mosquito *Anopheles gambiae*. *Insect Mol. Biol.* **14,** 575–589.

Blandin, S., Moita, L. F., Köcher, T., Wilm, M., Kafatos, F. C., and Levashina, E. A. (2002). Reverse genetics in the mosquito *Anopheles gambiae*: Targeted disruption of the Defensin gene. *EMBO Rep.* **3,** 852–856.

Blandin, S., Shiao, S. H., Moita, L. F., Janse, C. J., Waters, A. P., Kafatos, F. C., and Levashina, E. A. (2004). Complement-like protein TEP1 is a determinant of vectorial capacity in the malaria vector *Anopheles gambiae*. *Cell* **116,** 661–670.

Boisson, B., Jacques, J. C., Choumet, V., Martin, E., Xu, J., Vernick, K., and Bourgouin, C. (2006). Gene silencing in mosquito salivary glands by RNAi. *FEBS Lett.* **580,** 1988–1992.

Calvo, E., Dao, A., Pham, V. M., and Ribeiro, J. M. (2007). An insight into the sialome of *Anopheles funestus* reveals an emerging pattern in anopheline salivary protein families. *Insect Biochem. Mol. Biol.* **37,** 164–175.

Catteruccia, F., Nolan, T., Loukeris, T. G., Blass, C., Savakis, C., Kafatos, F. C., and Crisanti, A. (2000). Stable germline transformation of malaria mosquito *Anopheles stephensi. Nature* **405,** 959–962.

Catteruccia, F., Godfray, H. C., and Crisanti, A. (2003). Impact of genetic manipulation on the fitness of *Anopheles stephensi* mosquitoes. *Science* **299,** 1225–1227.

Catteruccia, F., Benton, J. P., and Crisanti, A. (2005). An Anopheles transgenic sexing strain for vector control. *Nat. Biotechnol.* **23,** 1414–1417.

Chen, H., Wang, J., Liang, P., Karsay-Klein, M., James, A. A., Brazeau, D., and Yan, G. (2004). Microarray analyses for identification of Plasmodium-refractoriness candidate genes in mosquitoes. *Genome* **47,** 1061–1070.

Chen, X., Marinotti, O., Whitman, L., Jasinskiene, N., Romans, P., and James, A. A. (2007). The *Anopheles gambiae* vitellogenin gene (*VGT2*) promoter directs persistent accumulation of a reporter gene product in transgenic *Anopheles stephensi* following multiple blood meals. *Am. J. Trop. Med. Hyg.* **76,** 1118–1124.

Cho, K. H., Cheon, H. M., Kokoza, V., and Raikhel, A. S. (2006). Regulatory region of the vitellogenin receptor gene sufficient for high-level, germ line cell-specific ovarian expression in transgenic *Aedes aegypti* mosquitoes. *Insect Biochem. Mol. Biol.* **36,** 273–281.

Christophides, G. K., Zdobnov, E., Barillas-Mury, C., Birney, E., Blandin, S., Blass, C., Brey, P. T., Collins, F. H., Danielli, A., Dimopoulos, G., Hetru, C., Hoa, N. T., *et al.* (2002). Immunity-related genes and gene families in *Anopheles gambiae*. *Science* **298,** 159–165.

Coates, C. J., Jasinskiene, N., Miyashiro, L., and James, A. A. (1998). *Mariner* transposition and transformation of the yellow fever mosquito, *Aedes aegypti. Proc. Natl Acad. Sci. USA* **95,** 3748–3751.

Coates, C. J., Jasinskiene, N., Morgan, D., Tosi, L. R., Beverley, S. M. and James, A. A. (2000). Purified mariner (Mos1) transposase catalyzes the integration of marked elements into the germ-line of the yellow fever mosquito, *Aedes aegypti*. *Insect Biochem. Mol. Biol.* **30,** 1003-1008.

Coates, C. J., Jasinskiene, N., Pott, G. B., and James, A. A. (1999). Promoter-directed expression of recombinant fire-fly luciferase in the salivary glands of Hermes-transformed *Aedes aegypti. Gene* **226,** 317–325.

Conde, R., Zamudio, F. Z., Rodríguez, M. H., and Possani, L. D. (2000). Scopine, an anti-malaria and anti-bacterial agent purified from scorpion venom. *FEBS Lett.* **471,** 165–168.

Curtis, C. F., and Graves, P. M. (1998). Methods for replacement of malaria vector populations. *J. Trop. Med. Hyg.* **91,** 43–48.

Dafaalla, T. H., Condon, G. C., Condon, K. C., Phillips, C. E., Morrison, N. I., Jin, L., Epton, M. J., Fu, G., and Alphey, L. (2006). Transposon-free insertions for insect genetic engineering. *Nat. Biotechnol.* **24,** 820–821.

de Lara Capurro, M., Coleman, J., Beerntsen, B. T., Myles, K. M., Olson, K. E., Rocha, E., Krettli, A. U., and James, A. A. (2000). Virus-expressed, recombinant single-chain antibody blocks sporozoite infection of salivary glands in *Plasmodium gallinaceum*-infected *Aedes aegypti. Am. J. Trop. Med. Hyg.* **62,** 427–433.

Dissanayake, S. N., Marinotti, O., Ribeiro, J. M., and James, A. A. (2006). angaGEDUCI: *Anopheles gambiae* gene expression database with integrated comparative algorithms for identifying conserved DNA motifs in promoter sequences. *BMC Genomics* **7,** 116.

Edwards, M. J., Lemos, F. J. A., Donnelly-Doman, M., and Jacobs-Lorena, M. (1997). Rapid induction by a blood meal of a carboxypeptidase gene in the gut of the mosquito *Anopheles gambiae. Insect Biochem. Mol. Biol.* **27,** 1063–1072.

Edwards, M. J., Moskalyk, L. A., Donelly-Doman, M., Vlaskova, M., Noriega, F. G., Walker, V. K., and Jacobs-Lorena, M. (2000). Characterization of a carboxypeptidase A gene from the mosquito, *Aedes aegypti*. *Insect Mol. Biol.* **9,** 33–38.

Enayati, A. A., Ranson, H., and Hemingway, J. (2005). Insect glutathione transferases and insecticide resistance. *Insect Mol. Biol.* **14,** 3–8.

Farkas, G., and Udvardy, A. (1992). Sequence of scs and scs′ *Drosophila* DNA fragments with boundary function in the control of gene expression. *Nucleic Acids Res.* **20,** 2604.

Fox, A. N., Pitts, R. J., Robertson, H. M., Carlson, J. R., and Zwiebel, L. J. (2001). Candidate odorant receptors from the malaria vector mosquito *Anopheles gambiae* and evidence of down-regulation in response to blood feeding. *Proc. Natl Acad. Sci. USA* **98,** 14693–14697.

Francischetti, I. M., Valenzuela, J. G., Pham, V. M., Garfield, M. K., and Ribeiro, J. M. (2002). Toward a catalog for the transcripts and proteins (sialome) from the salivary gland of the malaria vector *Anopheles gambiae*. *J. Exp. Biol.* **205,** 2429–2451.

Franz, G., and Savakis, C. (1991). Minos, a new transposable element from *Drosophila hydei*, is a member of the Tc1-like family of transposons. *Nucleic Acids Res.* **19,** 6646.

Franz, A. W., Sanchez-Vargas, I., Adelman, Z. N., Blair, C. D., Beaty, B. J., James, A. A., and Olson, K. E. (2006). Engineering RNA interference-based resistance to dengue virus type 2 in genetically-modified *Aedes aegypti*. *Proc. Natl Acad. Sci. USA* **103,** 4198–4203.

Fraser, M. J., Ciszczon, T., Elick, T., and Bauser, C. (1996). Precise excision of TTAA-specific lepidopteran transposons *piggyBac* (IFP2) and tagalong (TFP3) from the baculovirus genome in cell lines from two species of Lepidoptera. *Insect Mol. Biol.* **5,** 141–151.

Goltsev, Y., Hsiong, W., Lanzaro, R., and Levine, M. (2004). Different combinations of gap repressors for common stripes in Anopheles and Drosophila embryos. *Dev. Biol.* **275,** 435–446.

Gossen, M., and Bujard, H. (1992). Tight control of gene expression in mammalian cells by tetracycline-responsive promoters. *Proc. Natl Acad. Sci. USA* **89,** 5547–5551.

Gray, C. E., and Coates, C. J. (2004). High-level gene expression in *Aedes albopictus* cells using a baculovirus Hr3 enhancer and IE1 transactivator. *BMC Mol. Biol.* **5,** 8.

Grossman, G. L., Rafferty, C. S., Clayton, J. R., Stevens, T. K., Mukabayire, O., and Benedict, M. Q. (2001). Germline transformation of the malaria vector, *Anopheles gambiae*, with the *piggyBac* transposable element. *Insect Mol. Biol.* **10,** 597–604.

Groth, A. C., and Calos, M. P. (2004). Phage integrases: Biology and applications. *J. Mol. Biol.* **335,** 667–678.

Habtewold, T., Povelones, M., Blagborough, A. M., and Christophides, G. K. (2008). Transmission-blocking immunity in the malaria non-vector mosquito *Anopheles quadriannulatus* species A. *PLoS Pathog.* **4,** e1000070.

Handler, A. M., Zimowska, G. J., and Horn, C. (2004). Post-integration stabilization of a transposon vector by terminal sequence deletion in *Drosophila melanogaster*. *Nat. Biotechnol.* **22,** 1150–1154.

Hansen, I. A., Attardo, G. M., Park, J. H., Peng, Q., and Raikhel, A. S. (2004). Target of rapamycin-mediated amino acid signaling in mosquito anautogeny. *Proc. Natl Acad. Sci. USA* **101,** 10626–10631.

Holt, R. A., Subramanian, G. M., Halpern, A., Granger, G. G., Charlab, R., Nusskern, D. R., Wincker, P., Clark, A. G., Ribeiro, J. M. C., Wides, R., Salzberg, S. L., Loftus, B., *et al.* (2002). The genome sequence of the malaria mosquito *Anopheles gambiae*. *Science* **298,** 129–149.

Horn, C., and Wimmer, E. A. (2000). A versatile vector set for animal transgenesis. *Dev. Genes Evol.* **210,** 630–637.

Horn, C., Schmid, B. G., Pogoda, F. S., and Wimmer, E. A. (2002). Fluorescent transformation markers for insect transgenesis. *Insect Biochem. Mol. Biol.* **32,** 1221–1235.

Irvin, N., Hoddle, M. S., O'Brochta, D. A., Carey, B., and Atkinson, P. W. (2004). Assessing fitness costs for transgenic *Aedes aegypti* expressing the GFP marker and transposase genes. *Proc. Natl. Acad. Sci. USA* **101,** 891–896.

Isoe, J., Kunz, S., Manhart, C., Wells, M. A., and Miesfeld, R. L. (2007). Regulated expression of microinjected DNA in adult *Aedes aegypti* mosquitoes. *Insect Mol. Biol.* **16,** 83–92.

Ito, J., Ghosh, A., Moreira, L. A., Wimmer, E. A., and Jacobs-Lorena, M. (2002). Transgenic anopheline mosquitoes impaired in transmission of a malaria parasite. *Nature* **417,** 387–388.

Jasinskiene, N., Coates, C. J., Benedict, M. Q., Cornel, A. J., Rafferty, C. S., and James, A. A. (1998). State transformation of the yellow fever mosquito, *Aedes aegypti*, with the *Hermes* element from the horsefly. *Proc. Natl Acad. Sci. USA* **95,** 3743–3747.

Jasinskiene, N., Coates, C. J., and James, A. A. (2000). Structure of *Hermes* integrations in the germline of the yellow fever mosquito, *Aedes aegypti. Insect Mol. Biol.* **9,** 11–18.

Jasinskiene, N., Coates, C. J., Ashikyan, A., and James, A. A. (2003). High-efficiency, site-specific excision of a marker gene by the Phage P1 *cre–loxP* system in the yellow fever mosquito, *Aedes aegypti. Nucleic Acids Res.* **31**(22), e147.

Jasinskiene, N., Coleman, J., Ashikyan, A., Salampessy, M., Marinotti, O., and James, A. A. (2007). Genetic control of malaria parasite transmission: Threshold levels for infection in an avian model system. *Am. J. Trop. Med. Hyg.* **76,** 1072–1078.

Juhn, J., and James, A. A. (2006). *oskar* gene expression in the vector mosquitoes, *Anopheles gambiae* and *Aedes aegypti. Insect Mol. Biol.* **15,** 363–372.

Kato, N., Mueller, C. R., Fuchs, J. F., McElroy, K., Wessely, V., Higgs, S., and Christensen, B. M. (2008). Evaluation of the function of a type I peritrophic matrix as a physical barrier for midgut epithelium invasion by mosquito-borne pathogens in *Aedes aegypti. Vector Borne Zoonotic Dis.* **8,** 701–712.

Kellum, R., and Schedl, P. (1991). A position-effect assay for boundaries of higher order chromosomal domains. *Cell* **64,** 941–950.

Kim, W., Koo, H., Richman, A. M., Seeley, D., Vizioli, J., Klocko, A. D., and O'Brochta, D. A. (2004). Ectopic expression of a cecropin transgene in the human malaria vector mosquito *Anopheles gambiae* (Diptera: Culicidae): Effects on susceptibility to *Plasmodium. J. Med. Entomol.* **41,** 447–455.

Knols, B. G., Bossin, H. C., Mukabana, W. R., and Robinson, A. S. (2007). Transgenic mosquitoes and the fight against malaria: Managing technology push in a turbulent GMO world. *Am. J. Trop. Med. Hyg.* **77,** 232–242.

Kokoza, V., Ahmed, A., Cho, W. L., Jasinskiene, N., James, A. A., and Raikhel, A. (2000). Engineering blood meal-activated systemic immunity in the yellow fever mosquito, *Aedes aegypti. Proc. Natl Acad. Sci. USA* **97,** 9144–9149.

Kokoza, V., Ahmed, A., Wimmer, E. A., and Raikhel, A. S. (2001a). Efficient transformation of the yellow fever mosquito *Aedes aegypti* using the *piggyBac* transposable element vector pBac[3xP3-EGFP afm]. *Insect Mol. Biol.* **31,** 1137–1143.

Kokoza, V. A., Martin, D. M., Mienaltowski, M. J., Ahmed, A., Morton, C. M., Alexander, S., and Raikhel (2001b). Transcriptional regulation of the mosquito vitellogenin gene via a blood meal-triggered cascade. *Gene* **274,** 47–65.

Koutsos, A. C., Blass, C., Meister, S., Schmidt, S., MacCallum, R. M., Soares, M. B., Collins, F. H., Benes, V., Zdobnov, E., Kafatos, F. C., and Christophides, G. K. (2007). Life cycle transcriptome of the malaria mosquito *Anopheles gambiae* and comparison with the fruitfly Drosophila melanogaster. *Proc. Natl Acad. Sci. USA* **104,** 11304–11309.

Krzywinski, J., Grushko, O. G., and Besansky, N. J. (2006). Analysis of the complete mitochondrial DNA from *Anopheles funestus*: An improved dipteran mitochondrial genome annotation and a temporal dimension of mosquito evolution. *Mol. Phylogenet. Evol.* **39,** 417–423.

Kwon, H. W., Lu, T., Rützler, M., and Zwiebel, L. J. (2006). Olfactory responses in a gustatory organ of the malaria vector mosquito *Anopheles gambiae. Proc. Natl Acad. Sci. USA* **103,** 13526–13531.

Lobo, N. F., Clayton, J. R., Fraser, M. J., Kafatos, F. C., and Collins, F. H. (2006). High efficiency germ-line transformation of mosquitoes. *Nat. Protoc.* **1,** 1312–1317.

Lombardo, F., Nolan, T., Lycett, G., Lanfrancotti, A., Stich, N., Catteruccia, F., Louis, C., Coluzzi, M., and Arcà, B. (2005). An *Anopheles gambiae* salivary gland promoter analysis in *Drosophila melanogaster* and *Anopheles stephensi*. *Insect Mol. Biol.* **14,** 207–216.

Loukeris, T. G., Arcà, B., Livadaras, I., Dialektaki, G., and Savakis, C. (1995). Introduction of the transposable element *Minos* into the germ line of *Drosophila melanogaster*. *Proc. Natl Acad. Sci. USA* **92,** 9485–9489.

Lowenberger, C. A., Kamal, S., Chiles, J., Paskewitz, S., Bulet, P., Hoffmann, J. A., and Christensen, B. M. (1999). Mosquito–Plasmodium interactions in response to immune activation of the vector. *Exp. Parasitol.* **91,** 59–69.

Lycett, G. J., Kafatos, F. C., and Loukeris, T. G. (2004). Conditional expression in the malaria mosquito *Anopheles stephensi* with Tet-On and Tet-Off systems. *Genetics* **167,** 1781–1790.

Maragathavally, K. J., Kaminski, J. M., and Coates, C. J. (2006). Chimeric *Mos1* and *piggyBac* transposases result in site-directed integration. *FASEB J.* **20,** 1180–1182.

Marinotti, O., James, A. A., and Ribeiro, J. M. C. (1990). Diet and salivation in *Aedes aegypti* female mosquitoes. *J. Insect Physiol.* **36,** 545–548.

Marinotti, O., Nguyen, Q. K., Calvo, E., James, A. A., and Ribeiro, J. M. (2005). Microarray analysis of genes showing variable expression following a blood meal in *Anopheles gambiae*. *Insect Mol. Biol.* **14,** 365–373.

Marinotti, O., Calvo, E., Nguyen, Q. K., Dissanayake, S., Ribeiro, J. M., and James, A. A. (2006). Genome-wide analysis of gene expression in adult *Anopheles gambiae*. *Insect Mol. Biol* **15,** 1–12.

Marrelli, M. T., Moreira, C. K., Kelly, D., Alphey, L., and Jacobs-Lorena, M. (2006). Mosquito transgenesis: What is the fitness cost? *Trends Parasitol.* **22,** 197–202.

Marrelli, M. T., Li, C., Rasgon, J. L., and Jacobs-Lorena, M. (2007). Transgenic malaria-resistant mosquitoes have a fitness advantage when feeding on Plasmodium-infected blood. *Proc. Natl Acad. Sci. USA* **104,** 5580–5583.

McGrane, V., Carlson, J. O., Miller, B. R., and Beaty, B. J. (1988). Microinjection of DNA into *Aedes triseriatus* ova and detection of integration. *Am. J. Trop. Med. Hyg.* **39,** 502–510.

Medhora, M., Maruyama, K., and Hartl, D. L. (1991). Molecular and functional analysis of the *mariner* mutator element *MosI* in *Drosophila*. *Genetics* **128,** 311–318.

Meister, S., Koutsos, A. C., and Christophides, G. K. (2004). The Plasmodium parasite—A 'new' challenge for insect innate immunity. *Int. J. Parasitol.* **34,** 1473–1482.

Melo, A. C., Rützler, M., Pitts, R. J., and Zwiebel, L. J. (2004). Identification of a chemosensory receptor from the yellow fever mosquito, *Aedes aegypti*, that is highly conserved and expressed in olfactory and gustatory organs. *Chem. Senses* **29,** 403–440.

Mendes, A. M., Schlegelmilch, T., Cohuet, A., Awono-Ambene, P., De Iorio, M., Fontenille, D., Morlais, I., Christophides, G. K., Kafatos, F. C., and Vlachou, D. (2008). Conserved mosquito/parasite interactions affect development of *Plasmodium falciparum* in Africa. *PLoS Pathog.* **4,** e1000069.

Meredith, S. E. O., and James, A. A. (1990). Biotechnology as applied to vectors and vector control. *Ann. Parasitol. Hum. Comp.* **65,** 113–118.

Miller, L. H., Sakai, R. K., Romans, P., Gwadz, R. W., Kantoff, P., and Coon, H. G. (1987). Stable integration and expression of a bacterial gene in the mosquito *Anopheles gambiae*. *Science* **237,** 779–781.

Moreira, L. A., Edwards, M. J., Adhami, F., Jasinskiene, N., James, A. A., and Jacobs-Lorena, M. (2000). Robust gut-specific gene expression in transgenic *Aedes aegypti* mosquitoes. *Proc. Natl Acad. Sci. USA* **97,** 10895–10898.

Moreira, L. A., Ito, J., Ghosh, A., Devenport, M., Zieler, H., Abraham, E. G., Crisanti, A., Nolan, T., Catteruccia, F., and Jacobs-Lorena, M. (2002). Bee venom phospholipase inhibits malaria parasite development in transgenic mosquitoes. *J. Biol. Chem.* **277,** 40839–40843.

Moreira, L. A., Wang, J., Collins, F. H., and Jacobs-Lorena, M. (2004). Fitness of anopheline mosquitoes expressing transgenes that inhibit Plasmodium development. *Genetics* **166,** 1337–1341.

Morris, A. C., Eggleston, P., and Crampton, J. M. (1989). Genetic transformation of the mosquito *Aedes aegypti* by micro-injection of DNA. *Med. Vet. Entomol.* **3,** 1–7.

Morris, A. C., Schaub, T. L., and James, A. A. (1991). FLP-mediated recombination in the vector mosquito, *Aedes aegypti. Nucleic Acids Res.* **19,** 5895–5900.

Morris, A. C., Pott, G. B., Chen, J., and James, A. A. (1995). Expression of a promoter–reporter construct in differentiated adult salivary glands and embryos of the mosquito, *Aedes aegypti. Am. J. Trop. Med. Hyg.* **52,** 456–460.

Müller, H. M., Crampton, J. M., della Torre, A., Sinden, R., and Crisanti, A. (1993). Members of a trypsin gene family in *Anopheles gambiae* are induced in the gut by blood meal. *EMBO J.* **12,** 2891–2900.

Nene, V., Wortman, J. R., Lawson, D., Haas, B., Kodira, C., Tu, Z. J., Loftus, B., Xi, Z., Megy, K., Grabherr, M., Ren, Q., Zdobnov, E. M., *et al.* (2007). Genome sequence of *Aedes aegypti,* a major arbovirus vector. *Science* **316,** 1718–1723.

Nimmo, D. D., Alphey, L., Meredith, J. M., and Eggleston, P. (2006). High efficiency site-specific genetic engineering of the mosquito genome. *Insect Mol. Biol.* **15,** 129–136.

Nirmala, X., and James, A. A. (2003). Engineering plasmodium-refractory phenotypes in mosquitoes. *Trends Parasitol.* **19,** 384–387.

Nirmala, X., Marinotti, O., Sandoval, J. M., Phin, S., Gakhar, S., Jasinskiene, N., and James, A. A. (2006). Functional characterization of the promoter of the vitellogenin gene, *AsVg1,* of the malaria vector, *Anopheles stephensi. Insect Biochem. Mol. Biol.* **36,** 694–700.

Nolan, T., Bower, T. M., Brown, A. E., Crisanti, A., and Catteruccia, F. (2002). *piggyBac*-mediated germline transformation of the malaria mosquito *Anopheles stephensi* using the red fluorescent protein dsRED as a selectable marker. *J. Biol. Chem.* **277,** 8759–8762.

O'Brochta, D. A., Warren, W. D., Saville, K. J., and Atkinson, P. W. (1996). *Hermes,* a functional non-Drosophilid insect gene vector from *Musca domestica. Genetics* **142,** 907–914.

Park, J. H., Attardo, G. M., Hansen, I. A., and Raikhel, A. S. (2006). GATA factor translation is the final downstream step in the amino acid/target-of-rapamycin-mediated vitellogenin gene expression in the anautogenous mosquito *Aedes aegypti. J. Biol. Chem* **281,** 11167–11176.

Perera, O. P., Harrell, R. A., II, and Handler, A. M. (2002). Germ-line transformation of the South American malaria vector, Anopheles albimanus, with a *piggyBac*/EGFP transposon vector is routine and highly efficient. *Insect Mol. Biol.* **11,** 291–297.

Pham, D. Q., and Chavez, C. A. (2005). The ferritin light-chain homologue promoter in *Aedes aegypti. Insect Mol. Biol.* **14,** 263–270.

Pinkerton, A. C., Michel, K., O'Brochta, D. A., and Atkinson, P. W. (2000). Green fluorescent protein as a genetic marker in transgenic *Aedes aegypti. Insect Mol. Bio.l* **9,** 1–10.

Pitts, R. J., Fox, A. N., and Zwiebel, L. J. (2004). A highly conserved candidate chemoreceptor expressed in both olfactory and gustatory tissues in the malaria vector *Anopheles gambiae. Proc. Natl Acad. Sci. USA* **101,** 5058–5063.

Prévot, G. I., Laurent-Winter, C., Rodhain, F., and Bourgouin, C. (2003). Sex-specific and blood meal-induced proteins of *Anopheles gambiae* midguts: Analysis by two-dimensional gel electrophoresis. *Malar. J.* **2,** 1.

Ranson, H., and Hemingway, J. (2005). Mosquito glutathione transferases. *Methods Enzymol.* **401,** 226–241.

Ribeiro, J. M. (2003). A catalogue of *Anopheles gambiae* transcripts significantly more or less expressed following a blood meal. *Insect Biochem. Mol. Biol.* **33,** 865–882.

Ribeiro, J. M., and Francischetti, I. M. (2003). Role of arthropod saliva in blood feeding: Sialome and post-sialome perspectives. *Annu. Rev. Entomol.* **48,** 73–88.

Ribeiro, J. M., Charlab, R., Pham, V. M., Garfield, M., and Valenzuela, J. G. (2004). An insight into the salivary transcriptome and proteome of the adult female mosquito *Culex pipiens quinquefasciatus*. *Insect Biochem.* Mol. *Biol.* **34,** 543–563.

Ribeiro, J. M., Arcà, B., Lombardo, F., Calvo, E., Phan, V. M., Chandra, P. K., and Wikel, S. K. (2007). An annotated catalogue of salivary gland transcripts in the adult female mosquito, *Aedes aegypti. BMC Genomics* **8,** 6.

Rodrigues, F. G., Santos, M. N., de Carvalho, T. X., Rocha, B. C., Riehle, M. A., Pimenta, P. F., Abraham, E. G., Jacobs-Lorena, M., Alves de Brito, C. F., and Moreira, L. A. (2008). Expression of a mutated phospholipase A2 in transgenic *Aedes fluviatilis* mosquitoes impacts *Plasmodium gallinaceum* development. *Insect* Mol. *Biol.* **17,** 175–183.

Rützler, M., and Zwiebel, L. J. (2005). Molecular biology of insect olfaction: Recent progress and conceptual models. *J. Comp. Physiol. A Neuroethol. Sens. Neural. Behav. Physiol.* **191,** 777–790.

Sanchez-Vargas, I., Travanty, E. A., Keene, K. M., Franz, A. W., Beaty, B. J., Blair, C. D., and Olson, K. E. (2004). RNA interference, arthropod-borne viruses, and mosquitoes. *Insect* Mol. *Biol.* **102,** 65–74.

Sanders, H. R., Evans, A. M., Ross, L. S., and Gill, S. S. (2003). Blood meal induces global changes in midgut gene expression in the disease vector, *Aedes aegypti. Insect Biochem.* Mol. *Biol.* **33,** 1105–1122.

Sarkar, A., Atapattu, A., Belikoff, E. J., Heinrich, J. C., Li, X., Horn, C., Wimmer, E. A., and Scott, M. J. (2006). Insulated *piggyBac* vectors for insect transgenesis. *BMC Biotechnol.* **6,** 27.

Sato, K., Pellegrino, M., Nakagawa, T., Nakagawa, T., Vosshall, L. B., and Touhara, K. (2006). Insect olfactory receptors are heteromeric ligand-gated ion channels. *Nature* **452,** 1002–1006.

Scott, T. W. (2006). Fitness studies: Developing a consensus methodology. *In* "Bridging Laboratory and Field Research for Genetic Control of Disease" (B. G. J. Knols and C. Louis, eds.), pp. 171–181. Springer Science + Business Media, Wageningen UR Frontis Series. Springer, Dordrecht.

Sethuraman, N., Fraser, M. J., Jr, Eggleston, P., and O'Brochta, D. A. (2007). Post-integration stability of *piggyBac* in *Aedes aegypti. Insect Biochem.* Mol. *Biol* **37,** 941–951.

Severson, D. W., Meece, J. K., Lovin, D. D., Saha, G., and Morlais, I. (2002). Linkage map organization of expressed sequence tags and sequence tagged sites in the mosquito, *Aedes aegypti. Insect* Mol. *Biol* **11,** 371–378.

Shin, S. W., Kokoza, V. A., and Raikhel, A. S. (2003). Transgenesis and reverse genetics of mosquito innate immunity. *J. Exp. Biol.* **206,** 3835–3843.

Shotkoski, F., Morris, A. C., James, A. A., and ffrench-Constant, R. H. (1996). Functional analysis of the promoter of the γ-aminobutyric acid receptor gene of the mosquito, *Aedes aegypti. Gene* **168,** 127–133.

Sim, C., Hong, Y. S., Vanlandingham, D. L., Harker, B. W., Christophides, G. K., Kafatos, F. C., Higgs, S., and Collins, F. H. (2005). Modulation of *Anopheles gambiae* gene expression in response to o'nyong-nyong virus infection. *Insect* Mol. *Biol* **14,** 475–481.

Smith, R. C., Walter, M. F., Hice, R. H., O'Brochta, D. A., and Atkinson, P. W. (2007). Testis-specific expression of the beta2 tubulin promoter of *Aedes aegypti* and its application as a genetic sex-separation marker. *Insect* Mol. *Biol.* **16,** 61–71.

Srinivasan, P., Abraham, E. G., Ghosh, A. K., Valenzuela, J., Ribeiro, J. M., Dimopoulos, G., Kafatos, F. C., Adams, J. H., Fujioka, H., and Jacobs-Lorena, M. (2004). Analysis of the Plasmodium and Anopheles transcriptomes during oocyst differentiation. *J. Biol. Chem.* **279,** 5581–5587.

Thyagarajan, B., Olivares, E. C., Hollis, R. P., Ginsburg, D. S., and Calos, M. P. (2001). Site-specific genomic integration in mammalian cells mediated by phage phiC31 integrase. Mol. *Cell Biol.* **21,** 3926–3934.

Urlinger, S., Baron, U., Thellmann, M., Hasan, M. T., Bujard, H., and Hillen, W. (2000). Exploring the sequence space for tetracycline-dependent transcriptional activators: Novel mutations yield expanded range and sensitivity. *Proc. Natl Acad. Sci. USA* **97,** 7963–7968.

Valenzuela, J. G., Pham, V. M., Garfield, M. K., Francischetti, I. M. B., and Ribeiro, J. M. C. (2002) Toward a description of the sialome of the adult female mosquito *Aedes aegypti*. *Insect Biochem. Mol. Biol.* **32,** 1101–1122.

Vlachou, D., Schlegelmilch, T., Christophides, G. K., and Kafatos, F. C. (2005). Functional genomic analysis of midgut epithelial responses in *Anopheles* during *Plasmodium* invasion. *Curr. Biol.* **15,** 1185–1195.

Vontas, J., Blass, C., Koutsos, A. C., David, J. P., Kafatos, F. C., Louis, C., Hemingway, J., Christophides, G. K., and Ranson, H. (2005). Gene expression in insecticide resistant and susceptible *Anopheles gambiae* strains constitutively or after insecticide exposure. *Insect Mol. Biol.* **14,** 509–521.

Vontas, J., David, J. P., Nikou, D., Hemingway, J., Christophides, G. K., Louis, C., and Ranson, H. (2007). Transcriptional analysis of insecticide resistance in *Anopheles stephensi* using cross-species microarray hybridization. *Insect Mol. Biol.* **16,** 315–324.

Walker, W. B., Smith, E. M., Jan, T., and Zwiebel, L. J. (2008). A functional role for *Anopheles gambiae* Arrestin1 in olfactory signal transduction. *J. Insect Physiol.* **54,** 680–690.

Warr, E., Aguilar, R., Dong, Y., Mahairaki, V., and Dimopoulos, G. (2007). Spatial and sex-specific dissection of the *Anopheles gambiae* midgut transcriptome. *BMC Genomics* **8,** 37.

Waterhouse, R. M., Wyder, S., and Zdobnov, E. M. (2008). The *Aedes aegypti* genome: A comparative perspective. *Insect Mol. Biol.* **17,** 1–8.

Wetzel, C. H., Behrendt, H. J., Gisselmann, G., Störtkuhl, K. F., Hovemann, B., and Hatt, H. (2001). Functional expression and characterization of a Drosophila odorant receptor in a heterologous cell system. *Proc. Natl Acad. Sci. USA* **98,** 9377–9380.

Wilson, R., Orsetti, J., Klocko, A. D., Aluvihare, C., Peckham, E., Atkinson, P. W., Lehane, M. J., and O'Brochta, D. A. (2003). Post-integration behavior of a *Mos1* mariner gene vector in *Aedes aegypti*. *Insect Biochem. Mol. Biol.* **33,** 853–863.

Xi, Z., Ramirez, J. L., and Dimopoulos, G. (2008). The *Aedes aegypti* toll pathway controls dengue virus infection. *PLoS Pathog* **4,** e100098.

Xia, Y., Wang, G., Buscariollo, D., Pitts, R. J., Wenger, H., and Zwiebel, L. J. (2008). The molecular and cellular basis of olfactory-driven behavior in *Anopheles gambiae* larvae. *Proc. Natl Acad. Sci. USA* **105,** 6433–6438.

Yoshida, S., and Watanabe, H. (2006). Robust salivary gland-specific transgene expression in *Anopheles stephensi* mosquito. *Insect Mol. Biol.* **15,** 403–410.

Yoshida, S., Matsuoka, H., Luo, E., Iwai, K., Arai, M., Sinden, R. E., and Ishii, A. (1999). A single-chain antibody fragment specific for the *Plasmodium berghei* ookinete protein Pbs21 confers transmission blockade in the mosquito midgut. *Mol. Biochem. Parasitol.* **104,** 195–204.

Zdobnov, E. M., von Mering, C., Letunic, I., Torrents, D., Suyama, M., Copley, R. R., Christophides, G. K., Thomasova, D., Holt, R. A., Subramanian, G. M., Mueller, H. M., Dimopoulos, G., *et al.* (2002). Comparative genome and proteome analysis of *Anopheles gambiae* and *Drosophila melanogaster*. *Science* **298,** 149.

Zwiebel, L. J., and Takken, W. (2004). Olfactory regulation of mosquito–host interactions. *Insect Biochem. Mol. Biol.* **34,** 645–652.

3

Phylogeny of Tec Family Kinases: Identification of a Premetazoan Origin of Btk, Bmx, Itk, Tec, Txk, and the Btk Regulator SH3BP5

Csaba Ortutay,* Beston F. Nore,† Mauno Vihinen,*,‡ and C. I. Edvard Smith†

*Institute of Medical Technology, FI-33014 University of Tampere, Finland
†Department of Laboratory Medicine, Clinical Research Center, Karolinska Institutet, Karolinska University Hospital Huddinge, SE-141 86 Huddinge, Sweden
‡Tampere University Hospital, FI-33520 Tampere, Finland

Advances in Genetics, Vol. 64

0065-2660/08 $35.00
DOI: 10.1016/S0065-2660(08)00803-1

ABSTRACT

It is generally considered mammals and birds have five Tec family kinases (TFKs): Btk, Bmx (also known as Etk), Itk, Tec, and Txk (also known as Rlk). Here, we discuss the domains and their functions and regulation in TFKs. Over the last few years, a large number of genomes from various phyla have been sequenced making it possible to study evolutionary relationships at the molecular and sequence level. Using bioinformatics tools, we for the first time demonstrate that a TFK ancestor exists in the unicellular choanoflagellate *Monosiga brevicollis*, which is the closest known relative to metazoans with a sequenced genome. The analysis of the genomes for sponges, insects, hagfish, and frogs suggests that these species encode a single TFK. The insect form has a divergent and unique N-terminal region. Duplications generating the five members took place prior to the emergence of vertebrates. Fishes have two or three forms and the platypus, *Ornithorhynchus anatinus*, has four (lacks Txk). Thus, not all mammals have all five TFKs. The single identified TFK in frogs is an ortholog of Tec. Bmx seems to be unique to mammals and birds. SH3BP5 is a negative regulator of Btk. It is conserved in choanoflagellates and interestingly exists also in nematodes, which do not express TFKs, suggesting a broader function in addition to Btk regulation. The related SH3BP5-like protein is not found in Nematodes.

I. TYROSINE KINASES AND THE TEC FAMILY

A. Identification and characteristics of TFKs

In mammals, Tec family kinases (TFKs), Btk, Bmx (Etk), Itk, Tec, and Txk (Rlk), form the second largest family of cytoplasmic protein–tyrosine kinases (PTKs), the largest being related to Src, harboring nine Src family kinases (SFKs) (Caenepeel *et al.*, 2004; Quintaje and Orchard, 2008). We have used the official Human Genome Nomenclature Committee (HGNC) (http://www.genenames.org/) gene and protein names throughout the text. Human TFKs are in upper case and gene names in italics. The 2008 annotation lists 480 classical and 24 nonclassical protein kinases in man, out of which 90 are PTKs, while mice have

93 tyrosine kinases (Quintaje and Orchard, 2008). The protein kinases constitute one of the largest mammalian gene families comprising about 2% of all genes or about 10% of signaling functions coding genes.

With the exception of Txk, TFKs are characterized by an N-terminal pleckstrin homology (PH) domain, followed by a Tec homology (TH), Src homology (SH) -3, -2, and -1 (catalytic) domains (Smith *et al.*, 1994b; Vihinen *et al.*, 1994a). Txk instead has a cysteine-rich string, which, like PH domains, is required for temporary membrane attachment. PH domains in TFKs are also involved in binding to heterotrimeric G proteins and protein serine/threonine kinases (PSKs).

The TH domain consists of an N-terminal Zn^{2+}-binding Btk motif (Hyvönen and Saraste, 1997; Vihinen *et al.*, 1994a, 1997a) and one or two proline-rich motifs (Smith *et al.*, 2001; Vihinen *et al.*, 1994a, 1997a). Bmx lacks the typical proline-rich region (PRR) of the TH domain and also has an altered SH3 domain (Smith *et al.*, 2001; Tamagnone *et al.*, 1994). The PRRs participate in inter- and intramolecular SH3 domain binding. The SH2 and SH3 domains are docking modules, which bind to polyproline helices and phosphotyrosines, respectively. The kinase domain is the only catalytic entity in TFKs.

The mammalian TFKs were cloned during 1990–1995 (Haire *et al.*, 1994; Heyeck and Berg, 1993; Hu *et al.*, 1995; Mano *et al.*, 1990; Robinson *et al.*, 1996; Siliciano *et al.*, 1992; Tamagnone *et al.*, 1994; Tsukada *et al.*, 1993; Vetrie *et al.*, 1993; Yamada *et al.*, 1993) and immediately received wide interest, especially owing to the fact that *BTK* mutations cause an X-linked form of B lymphocyte deficiency (X-linked agammaglobulinemia, XLA) in man (Lindvall *et al.*, 2005; Tsukada *et al.*, 1993; Väliaho *et al.*, 2006; Vetrie *et al.*, 1993; Vihinen *et al.*, 1995a) and X-linked immunodeficiency (XID) in mice (Rawlings *et al.*, 1993; Thomas *et al.*, 1993). Btk is expressed in lymphoid cells but absent from T cells and plasma cells (Smith *et al.*, 1994a).

Disease-related mutations affecting the corresponding enzyme in T lymphocytes, ITK, has to date not been reported, but this gene resides on an autosome making any loss-of-function phenotype considerably less common than that observed for BTK. However, Itk plays an essential role in T-lymphocyte development as shown by knocking out the gene in mice (Liao and Littman, 1995). The gene was initially identified using degenerate primers to amplify cDNA from an IL-2-dependent mouse T-cell line (Siliciano *et al.*, 1992) or from neonatal mouse thymus (Heyeck and Berg, 1993).

Tec was cloned from a hepatic carcinoma (Mano *et al.*, 1990), but was later found to be expressed in several tissues, including B lymphocytes (Mano *et al.*, 1993). The phenotype of mice with Btk deficiency is much milder than the one seen in humans. While Tec is also expressed in human B lymphocytes, in mice the generation of double knockouts for Btk and Tec causes an XLA-like phenotype, whereas Tec single-knockout mice do not have an overt phenotype (Ellmeier *et al.*, 2000). An osteoclast defect has been reported both in isolated Btk deficiency (Lee *et al.*, 2008) and in combined Btk/Tec deficiency (Shinohara *et al.*, 2008). More subtle effects of Tec have been recognized in platelets (Atkinson *et al.*, 2003; Crosby and Poole, 2002; Oda *et al.*, 2000), erythroid (Schmidt *et al.*, 2004a; van Dijk *et al.*, 2000), and phagocytic cells (Jongstra-Bilen *et al.*, 2008; Melcher *et al.*, 2008). In these cells, Tec is important for signaling through various receptors (Atkinson *et al.*, 2003; Crosby and Poole, 2002; Oda *et al.*, 2000; Schmidt *et al.*, 2004b; van Dijk *et al.*, 2000), and lack of Tec resulted in increased levels of caspases (Melcher *et al.*, 2008). Furthermore, Tec showed a unique late effect in the phagocytic process (Jongstra-Bilen *et al.*, 2008).

Bmx was originally identified from a bone marrow-derived cDNA library (Tamagnone *et al.*, 1994) and was later found to be expressed mainly in endothelial cells as well as in prostate tumors (Ekman *et al.*, 1997; Robinson *et al.*, 1996). Loss-of-function mutations in humans have not been detected, which is somewhat unexpected owing to its X-chromosomal location and the viability of the mouse knockout. Such mice are characterized by defects in arteriogenesis and angiogenesis (He *et al.*, 2006; Rajantie *et al.*, 2001).

Txk was first identified from human peripheral blood and murine thymus cDNA libraries, respectively (Haire *et al.*, 1994; Hu *et al.*, 1995), and was later found to be mainly involved in T-lymphocyte development, since Itk$^{-/-}$ and Txk$^{-/-}$ mice have a more severe phenotype as compared to an Itk$^{-/-}$ single defect (Broussard *et al.*, 2006; Gomez-Rodriguez *et al.*, 2007; Schaeffer *et al.*, 1999). Recently a role for Txk has also been described in NKT-cell development (Felices and Berg, 2008).

B. Biological functions of TFKs

The TFKs have been implicated as pivotal components of signaling pathways downstream of extracellular receptor stimuli, such as lymphocyte antigen receptors (Felices *et al.*, 2007; Lindvall *et al.*, 2005; Schmidt *et al.*, 2004a,b). A functional defect of Itk and Btk kinases affect both innate and adaptive immunity in T cells (Berg *et al.*, 2005) and B cells (Brunner *et al.*, 2005; Hasan *et al.*, 2008; Mansell *et al.*, 2006), respectively. Activation of TFKs is a two-step event, which requires phosphorylation by a SFK member, and translocation to the plasma membrane, mediated by the PH domain (Lewis *et al.*, 2001; Nore *et al.*, 2000; Varnai *et al.*, 2005). It is not known which of these comes first, but presumably membrane translocation is the initial event, since Btk mutants lacking membrane-binding activity are not phosphorylated. Moreover, SFKs are known to mainly reside in the membrane (Ingley, 2008) In the case of Txk, which lacks PH domain, N-terminal palmitoylated cysteine-string motif is responsible for membrane colocalization.

Btk seems to have a dual role in apoptosis, under certain conditions being protective, while in other cellular contexts it instead induces apoptosis (Islam and Smith, 2000; Uckun, 1998). Bmx was reported to protect cells from apoptosis (Xue *et al.*, 1999). The antiapoptotic role of TFKs may involve AP-1 signaling (Altman *et al.*, 2004). Another potentially antiapoptotic pathway is through NF-κB, which is downstream of Btk (Bajpai *et al.*, 2000; Petro *et al.*, 2000). Recently was shown that NF-κB acts on the Btk promoter region by feedback activation (Yu *et al.*, 2008).

One common phenomenon in TFK-mediated activation is that these kinases modulate actin polymerization and dynamics, which play a central role in cytoskeleton processes including cell division, motility, cell shape, and chemotaxis (Finkelstein and Schwartzberg, 2004; Gomez-Rodriguez *et al.*, 2007; Nore *et al.*, 2000). Btk29A in *Drosophila* is essential for head involution during embryogenesis and also for ring canal growth (Guarnieri *et al.*, 1998; Roulier *et al.*, 1998). Recent studies show that Btk29A controls actin remodeling (Chandrasekaran and Beckendorf, 2005) and microfilament contraction during embryonic cellularization (Thomas and Wieschaus, 2004). It is also of interest to note that the phenotype of flies lacking only the full-length isoform of Btk29A is not lethal (Baba *et al.*, 1999). Instead, the development of male genitals and longevity are affected. The phenotype of TFK-deficient, more primitive species, Porifera (sponge) and choanoflagellate, is not known.

II. AIM OF THIS REVIEW

A few reports have discussed the evolution of TFKs (Baba *et al.*, 1999; Cetkovic *et al.*, 2004; Guarneri *et al.*, 1998; Haire *et al.*, 1997, 1998; Nars and Vihinen, 2001; Roulier *et al.*, 1998; Smith *et al.*, 2001), but there are no recent publications related to this topic. Owing to the many genomes characterized over the last years, it seemed timely to compile and analyze the existing data.

SH3BP5 (also called Sab and in flies Parcas) has been identified as a negative regulator of at least Btk in both mammals and *Drosophila melanogaster* (Hamada *et al.*, 2005; Matsushita *et al.*, 1998; Sinka *et al.*, 2002). For this reason the phylogeny of this regulator was also investigated. The emergence and functions of TFKs as well as SH3BP5 are discussed from an evolutionary point of view.

III. THE FUNCTION OF INDIVIDUAL DOMAINS IN TFKs

A. Function of the SH3–SH2–kinase domain complex

Apart from TFKs, the SH3–SH2–kinase domain organization is common to most cytoplasmic tyrosine kinases, namely the Src, Brk/Srm/Frk, Csk, and the Abl families (Mattsson *et al.*, 1996; Serfas and Tyner, 2003), while the Syk/Zap family lacks an SH3 domain and has two SH2 domains and the Fes/Fer family has a single SH2 module located N-terminally to the kinase domain. Jak, Fak, and Ack cytoplasmic tyrosine kinases lack SH3 and SH2 domains. New kinases have emerged by simultaneous duplication of all the three domains together (Nars and Vihinen, 2001).

Owing to that the function of SH3, SH2, and kinase (SH1) domains has been widely investigated and reviewed (Lappalainen *et al.*, 2008; Pawson and Scott, 2005; Pawson *et al.*, 2001; Seet *et al.*, 2006; Williams and Zvelebil, 2004) and since these entities are not unique for TFKs, we will not discuss them in detail, but simply make a few remarks regarding TFK-related properties.

To date there are structures for the SH3 domain of Btk (Hansson *et al.*, 1998), Itk (Andreotti *et al.*, 1997; Laederach *et al.*, 2002, 2003; Severin *et al.*, 2008), and Tec (Pursglove *et al.*, 2002). The thermal unfolding pattern of different TFK SH3 domains has also been analyzed (Knapp *et al.*, 1998). In contrast to most other SH3 domains, the TFK family seems to be regulated by tyrosine phosphorylation as demonstrated for Btk (Park *et al.*, 1996) and Itk (Hao and August, 2002; Wilcox and Berg, 2003) as well as Tec and Bmx (Nore *et al.*, 2003). In Btk, Y223 is in addition to autophosphorylation (Park *et al.*, 1996) target also for Abl (Bäckesjö *et al.*, 2002).

Since SH3 domains bind proline-rich stretches, it is interesting to note that Btk, Itk, and Tec have conserved PRRs in the adjacent TH domain that could bind to the SH3 domain. The Btk motif structure has been solved for BTK together with PH domain (Hyvönen and Saraste, 1997), and the entire TH domain for BMX and ITK [Protein Data Bank (PDB) codes 2ys2 and 2e61, respectively]. Bmx lacks both a typical SH3 domain and the PRR. The fact that both stretches are altered in Bmx is compatible with a functional relationship between a proline-rich, or a polyproline-like region, and the SH3 domain, as manifested in Btk (Hansson *et al.*, 2001a,b; Laederach *et al.*, 2002), Itk (Andreotti *et al.*, 1997; Brazin *et al.*, 2000; Hao and August, 2002; Márquez *et al.*, 2003) as well as in SFKs (Moarefi *et al.*, 1997; Sicheri *et al.*, 1997; Wang *et al.*, 2007; Williams *et al.*, 1997; Xu *et al.*, 1997). The PH and TH domains are missing from Txk.

In the case of Btk there are two PRRs in the TH domain (Smith *et al.*, 2001; Vihinen *et al.*, 1994a) making it possible to form multiple interactions, both intra- and intermolecular in origin (Hansson *et al.*, 2001a,b; Laederach *et al.*, 2002; Okoh and Vihinen, 2002). This is not the case for Itk, which has a single proline-rich stretch, while for Txk the interaction is mainly intermolecular due to the short connecting linker (Laederach *et al.*, 2002, 2003). However, most of these studies were performed using constructs expressing only the PRR in combination with the SH3 domain. In the single study where a full-length TFK was studied, there was no indication of TH–SH3 interactions (Márquez *et al.*, 2003). As many multidomain proteins, including TFKs, have several overall protein folds depending, for example, on posttranslational modifications such as phosphorylation, or ligand binding, the intramolecular interactions vary. Kinases are known to have several such conformational changes due to active site loop phosphorylation, upper lobe twist because of ATP binding, and C-terminal phosphorylation in regulation of SFKs etc. Recently, Joseph *et al.* (2007a) presented evidence for an intramolecular *cis* mechanism for the phosphorylation of tyrosine 180 in the Itk SH3 domain.

NMR structural studies combined with mutational analysis demonstrated a proline-dependent conformational switch within the Itk SH2 domain. This switch regulates substrate recognition (Brazin *et al.*, 2000). *Cis–trans* isomerization of a single prolyl-imide bond (D286–P287) within the SH2 domain influenced substrate recognition (Breheny *et al.*, 2003; Mallis *et al.*, 2002). In Btk the corresponding protein has only *trans* conformation (Huang *et al.*, 2006). Originally, the BTK SH2 domain was modeled (Vihinen *et al.*, 1994b) and used to analyze structure–function relationships along with biophysical methods for selected mutants (Mattsson *et al.*, 2000). The structural consequences of all reported SH2 domain mutations in altogether 10 diseases have been investigated with bioinformatics methods (Lappalainen *et al.*, 2008). *Cis–trans* isomerization instead seems to take place within the Btk PH domain (Yu *et al.*, 2006) as described below.

The linker between the SH2 and kinase domain in Itk was recently shown to positively regulate catalysis (Joseph *et al.*, 2007b) and the SH2 domain seems to be involved in substrate binding, not only in Itk, but also in Btk and Tec (Joseph *et al.*, 2007c).

The kinase domain structure has been solved for BTK (Mao *et al.*, 2001) and ITK (Brown *et al.*, 2004). These domains are highly conserved in all TFK sequences.

B. Regulation of PH domain binding by phosphoinositide 3-kinase (PI3K)

The TFKs are the only tyrosine kinases having a PH domain. Likewise, the TH domain is unique to this family. While functional analyses of individual TFKs have demonstrated unique features, they also have many common characteristics. Activation of PI3K generates phosphatidylinositol-3,4,5-trisphosphate

(PIP_3) serving as a membrane docking site for the PH domain of TFKs. PIP_3 is the most negatively charged plasma membrane lipid, concentration of which can increase 40-fold within seconds after PI3K activation (Stephens *et al.*, 1993). PIP_3 binding characterizes the PH domain of Btk (Manna *et al.*, 2007; Nore *et al.*, 2000; Rameh *et al.*, 1997; Salim *et al.*, 1996; Watanabe *et al.*, 2003), Bmx (Ekman *et al.*, 2000; Jiang *et al.*, 2007; Qiu *et al.*, 1998), Itk (August *et al.*, 1997; Huang *et al.*, 2007; Lu *et al.*, 1998), as well as Tec (Kane and Watkins, 2005; Lachance *et al.*, 2002; Tomlinson *et al.*, 2004). TFKs have widely varying binding specificities and affinities for inositol compounds (Kojima *et al.*, 1997). The differences in the binding have been investigated by modeling the structures for Bmx, Itk, and Tec PH domains (Okoh and Vihinen, 1999).

A large set of phosphoinositide-binding PH domains in various species were identified by combining bioinformatics with experimental studies (Park *et al.*, 2008). This study confirmed the PIP_3-inducible binding of Btk and Tec. Altogether, 40 PI3K-regulated PH domain-containing proteins were identified in vertebrates, four in *D. melanogaster*, one of which was Btk29A, but none in yeast. Amino acids across the whole PH domain were found to contribute to PIP_3 binding. The evolutionary interpretations from the study were that PIP_3 regulation of PH domains has evolved several times as independent events.

PI3K can be subdivided into three classes: IA, IB; II; and III (Vanhaesebroeck *et al.*, 1997). From an evolutionary point of view PI3K has been identified both in yeast and in plants (Herman *et al.*, 1992; Hong and Verma, 1994; Welters *et al.*, 1994). As PIP_3 can be generated via stimulation of many different receptors, including immunoreceptors, G protein-coupled receptors, as well as membrane-spanning tyrosine kinases, PI3K-induced activation of TFKs can be achieved through multiple pathways. At physiological concentrations, IP_4 enhances the binding of Itk's PH domain to PIP_3 (Huang *et al.*, 2007). In the case of Btk, its selective interaction with particular phosphoinositides has been addressed using biochemical and cell-biological methods (Hamman *et al.*, 2002; Nore *et al.*, 2000; Rameh *et al.*, 1997; Saito *et al.*, 2001; Salim *et al.*, 1996; Varnai *et al.*, 2005) as well as structure determination (Baraldi *et al.*, 1999).

C. Regulation of TFKs through the PH domain by serine/threonine kinases

The PH domain also serves as an important region for the regulation of TFKs by serine/threonine kinases. However, the situation is quite complex with differential effects among family members. The activation of protein kinase C negatively regulates the activity of Btk (Yao *et al.*, 1994). These and other authors demonstrated that PKCβI constitutively interacts with Btk *in vivo* and that both Ca^{2+}-dependent and -independent forms of PKC could bind to Btk (Johannes *et al.*, 1999; Kawakami *et al.*, 2000; Yao *et al.*, 1994). Btk also serves as a substrate for

PKC and its enzymatic activity is downregulated by PKC-mediated phosphorylation. However, more recent studies have reported that the key regulatory site, S180, is in fact in the TH domain (Kang *et al.*, 2001; Venkataraman *et al.*, 2006).

In platelets, PKCq activates Btk, while Btk negatively regulates PKCq (Crosby and Poole, 2002). Although, PKCβI and -II serve as negative regulators of Btk, the deletion of the gene encoding PKCβI and -II in mice causes a phenocopy of Btk deficiency (Leitges *et al.*, 1996). In Itk, the situation may be different, since PKC seems to activate it (Kawakami *et al.*, 1996).

Tec binds constitutively to PKCθ through its PH domain (Altman *et al.*, 2004). The Bmx-induced DNA binding of Stat1 is selectively inhibited by PKCδ. The coexpression of Bmx with PKCδ-induced phosphorylation of this isoform of PKC (Saharinen *et al.*, 1997). The interaction between PH domains and PKC seems to extend to at least some other PH domains, since the PH domain of G protein-coupled receptor kinase-2, GRK2, binds to PKC and affects the activity of this kinase (Yang *et al.*, 2003).

Evolutionarily, PKCs are ancient and found also in plants (Bögre *et al.*, 2003; Zegzouti *et al.*, 2006) and in yeast (Levin *et al.*, 1990). The same is true for phospholipase C, which appeared more than 1000 million years ago (Hirayama *et al.*, 1995; Koyanagi *et al.*, 1998; Tasma *et al.*, 2008). PLCγ2 acts upstream of PKCs and serves as a substrate for Btk, which phosphorylates two tyrosines in a linker between the SH2 and SH3 domains (Humphries *et al.*, 2004; Kim *et al.*, 2004).

Recently, the importance of serines and PSK phosphorylation sites has been revealed from another direction. Peptidylprolyl-isomerase Pin1 targets in BTK two PH domain dipeptides, S21–P22 and S115–P116, and acts as a negative regulator. Pin1 affects the most N-terminal site in mitosis and S115–P116 during the interphase (Yu *et al.*, 2006). Corresponding serine/threonine kinases or phosphatases have not been identified until now. Pin1 appears also in plants (Yao *et al.*, 2001) and in yeast (Behrsin *et al.*, 2007; Hanes *et al.*, 1989; Lu *et al.*, 1996).

D. Function of the TH domain

Another conserved, characteristic region of TFKs is the TH domain, and especially its N-terminal Btk motif (Smith *et al.*, 1994b; Vihinen *et al.*, 1994a, 1997a). While the PRR differs among TFKs, the Btk motif is highly conserved. Its only known function is to stabilize the PH domain through the conformational interaction with a Zn^{2+} ion, as demonstrated for Btk (Baraldi *et al.*, 1999; Hyvönen and Saraste, 1997). The entire TH domain is unique for TFKs. The Btk motif appears also in some other proteins. In the in-depth report of TH domain identification (Vihinen *et al.*, 1994a) an unidentified partial protein was detected, which turned out to be Ras GTPase-activating protein (GAP) (Vihinen *et al.*, 1997a).

As discussed above, the PRR part of the TH domain can interact with the SH3 domain in an inter- or intramolecular manner. Whether additional functions exist is not known. Missense mutations affecting patients with XLA result in a very unstable protein, which readily is degraded (Vihinen *et al.*, 1997a). Structure for separately expressed domains in Bmx and Itk with Zn^{2+} have been solved.

E. Regulation of Btk through SH3BP5

SH3BP5 (Sab) was originally identified in mammals as a novel adaptor protein, which binds to the SH3 domain of Btk with high preference (Matsushita *et al.*, 1998). SH3BP5 also controls negatively B-cell antigen-mediated signaling (Yamadori *et al.*, 1999). *D. melanogaster* SH3BP5 (Sinka *et al.*, 2002), also denoted Parcas, was demonstrated in a genetic screen to act as a negative regulator of Btk29A (Hamada *et al.*, 2005).

SH3BP5 also interacts with c-Jun N-terminal kinase (JNK) (Wiltshire *et al.*, 2002). As with c-Jun, the JNK interaction is mediated through its putative mitogen-activated protein kinase interaction motifs (KIMs) (Wiltshire *et al.*, 2002, 2004). Active JNK and phosphorylated Sab are colocalized and associated with mitochondria (Wiltshire *et al.*, 2002). These findings suggest a role in crosstalk between Btk and JNK signaling pathways (Wiltshire *et al.*, 2002, 2004).

A SH3BP5-like (SH3BP5L) protein shares sequence similarity with SH3BP5 (Strausberg *et al.*, 2002). No biological function has been characterized for it yet.

F. Mutations in X-linked agammaglobulinemia as a tool to study the function of residues

XLA, being the protein kinase with the largest number of disease-related mutation among all human protein kinases (Ortutay *et al.*, 2005), has allowed us and others to investigate the significance and structure–function relationships of several regions and amino acids in TFKs and other kinases. Missense and other in-frame mutations have been instrumental in these studies. Due to the conservation of TFKs along the entire sequence it is relatively easy to interpolate information for one family to others as in the case of JAK3 (Notarangelo *et al.*, 2001; Vihinen *et al.*, 2000). The importance of such data is not limited to TFKs, since when the naturally occurring, disease-causing mutation R28C in the Btk PH domain, in both in man and mice, was introduced to the Akt kinase, a stable but inactive protein was obtained (Lehnes *et al.*, 2007; Sable *et al.*, 1998; Stoica *et al.*, 2003).

Since 1995 mutations in the *BTK* gene have been collected. The BTKbase (http://bioinf.uta.fi/BTKbase) database is freely available and has served as a model for some 130 additional mutation databases, mainly for immunodeficiencies (Piirilä *et al.*, 2006). BTKbase has constantly grown from 188 cases to the current number of 1096 patients (Lindvall *et al.*, 2005; Väliaho *et al.*, 2006; Vihinen *et al.*, 1995a,b, 1996, 1997b, 1998, 1999, 2001). Experimental and modeled structures for BTK domains have extensively been used to explain protein structure–function relationships and consequences of mutations (Holinski-Feder *et al.*, 1998; Jin *et al.*, 1995; Korpi *et al.*, 2000; Lindvall *et al.*, 2005; Maniar *et al.*, 1995; Mao *et al.*, 2001; Mattsson *et al.*, 2000; Okoh *et al.*, 2002; Speletas *et al.*, 2001; Väliaho *et al.*, 2006; Vihinen *et al.*, 1994b,c, 1995b; Vorechovsky *et al.*, 1995, 1997; Zhu *et al.*, 1994). These studies have allowed us to provide putative functional and/or structural explanation for all XLA-causing mutations. The quality of the structural predictions was retrospectively assessed and found to be very good (Khan and Vihinen, submitted for publication).

Mutations have also been detected in the BTK promoter region affecting a highly conserved binding site for Ets family transcription factors (Holinski-Feder *et al.*, 1998). Btk has been reported to activate NF-κB signaling (Bajpai *et al.*, 2000; Petro *et al.*, 2000). Recently Btk was shown to autoregulate its promoter in a positive fashion (Yu *et al.*, 2008). Interestingly, also Bmx, Itk, and Tec seem to be positively regulated by NF-κB (Yu *et al.*, 2008, manuscript in preparation).

IV. THE ANCESTRY OF TFKs

A. Collecting the sequences

Multiple blastp (Altschul *et al.*, 1997) searches were made in an iterative manner against GenBank nonredundant protein database to identify TFK sequences. In the first step, human TFK protein sequences were used as query and the results were examined by distant tree representation with the online tool provided by the National Center for Biotechnology Information (NCBI). All the detected sequences were carefully checked and only full-length entries were accepted. When sequence variants appeared, they were aligned to the query sequence and the one closest to the query with the longest sequence was selected. From each search only those sequences, which were closest to the human query were included to the result set, that is, only the orthologs were included. Sequences for which the orthology was not so clear were collected separately. Additional search was made by using the fruit fly Btk29A sequence as the query.

TFK members were identified also from some incomplete genome datasets. Unannotated TFK sequences were searched from *Xenopus laevis* (Bowes *et al.*, 2008), *Xenopus tropicalis* (Bowes *et al.*, 2008), *Takifugu rubripes* (Aparicio *et al.*, 2002), and *Danio rerio* (zebrafish) (Sprague *et al.*, 2008) genomes. Several tblastn (Altschul *et al.*, 1997) searches were performed using TFK protein sequences from *Gallus gallus* (chicken) as queries. An obviously missing Btk sequence from *Macaca mulatta* was identified by using a tblastn search against Macaca mRNA database (Gibbs *et al.*, 2007). A likely mistranslated mRNA was detected with a frameshift caused by a possible sequencing error. The sequence was reconstructed by translating the mRNA in six reading frames and by aligning the protein sequences with the human BTK protein sequence. Partial sequences, such as those for Tec and Txk in *Sus scrofa* (wild boar), were excluded.

B. Aligning the sequences

The identified TFK sequences were divided into seven groups: the chordate-specific Bmx, Btk, Itk, Tec, and Txk groups and the insect-specific Btk29A-related sequences. The seventh group is for three sequences equally distant from all the others. This outgroup contained sequences BAD52302 from *Eptatretus burgeri* (hagfish), XP_001745298 from *Monosiga brevicollis* (choanoflagellate), and AAP82507 from *Suberites domuncula* (sponge). The six protein groups were aligned individually and finally combined with the outgroup sequences. In this way the internal conservation of the protein groups was better preserved in the final alignment.

C. Phylogenetic analysis

Based on the multiple sequence alignment, a bootstrap analysis was performed using maximum parsimony as criteria for searching the optimal tree (Fig. 3.1). The six protein groups are clearly separated on the tree, and we can draw the phylogeny of these proteins as follows. The ancestor of all the TFKs was present in early eukaryotes prior to the formation of metazoans. The sequences from *S. domuncula* and *M. brevicollis* are orthologs of the ancestor. After the divergence of deuterostomia and protostomia the descendants of the ancestor further diverged. In protostomia, now insects, TFKs developed in the form of the Btk29A protein group.

In deuterostomia a descendant of the single gene became the ancestor for the five chordata-specific protein groups. The TFK in *E. brugeri* is a direct ascendant to that. After the formation of craniata, but before the formation of vertebrata, the ancestor went through multiple duplications. First, it was divided into the Btk/Bmx and Tec/Txk/Itk groups. Then both groups duplicated until all the five protein groups appeared. These events took place before the emergence of vertebrates. The lack of sequences in fishes and some other genomes is likely due to deletion events rather than duplications after the emergence of the vertebrates, since all the sequences within the groups are more similar to each other than to any of the fish or frog sequences.

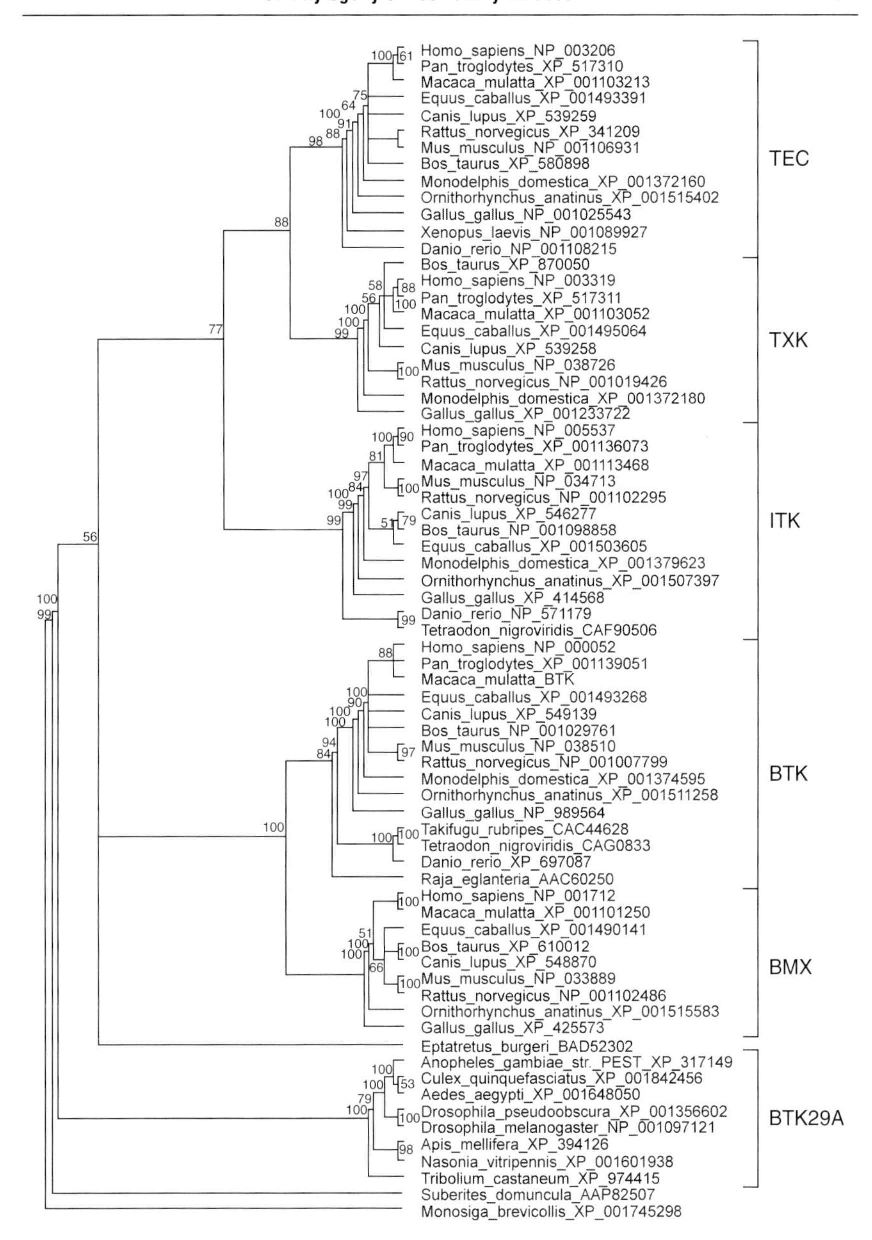

Figure 3.1. Phylogenetic relationship of the TFK sequences. The six major protein groups can be clearly distinguished. Bootstrap analysis was carried out using maximum parsimony criteria and 100 replicates with PAUP* (Swofford, 2003). Bootstrap values are shown at the nodes. Sequence labels contain name of the species and NCBI Entrez accession number for the protein sequence. Btk sequence for *Macaca mulatta* was reconstructed by transcribing the mRNA record XR_011285.1 and correcting the disrupted frameshift.

Another view on evolution is based upon the analysis of genomic organization. Recently the amphioxus, *Branchiostoma floridae*, genome was published (Putnam *et al.*, 2008). The authors partly reconstructed the genomic organization of the last common chordate ancestor and described two genome-wide duplications and subsequent reorganizations in the vertebrate lineage. Interestingly, number 8 of the 17 reconstructed ancestral chordate linkage groups contains regions corresponding to the location of all TFKs in the human genome.

Since we identified only a single frog TFK, the genomic region was analyzed. According to Xenbase (http://www.xenbase.org/), *Cyfip2* and *Med7* genes are both located in the vicinity of the *Tec* gene. This is surprising since both these genes are in close proximity to the *Itk* gene in zebrafish, mouse and man. The *Itk* and *Tec* genes in these three species are on different chromosomes. The *Txk* gene, which is absent from the zebrafish, is in very close proximity to the *Tec* gene in humans and mice. However, there is no doubt from the sequence alignment that the frog TFK should be classified as Tec. It is possible that a recombination event in an ancestor have transferred the *Tec* gene from its original location into the position of an *Itk* homolog, which was simultaneously lost. An alternative explanation is that the gene has evolved so that it is currently more closer to *Tec* than to *Itk*. The *X. laevis* Tec has similarity to Tec from other species throughout the sequence, suggesting that if a recombination occurred, the entire gene was replaced.

D. The origin of the Btk-specific PH domain loop in amniotes

Many Btk sequences contain a characteristic loop of 20 residues between amino acids 78 and 98 (numbering according to BTK). It is present only in the mammalian and bird Btk orthologs, not in the other TFKs. Exon 4 in these genes encodes residues 81–103 (human BTK numbering). We assume that the loop emerged by the insertion of exon 4 into the *Btk* gene. This insertion was tolerated, because the loop is on the surface of the protein (Fig. 3.2). The termini are close to each other in the three-dimensional space as in many protein domains, which appear in several proteins surrounded by different structural regions. In the loop region, there are in BTKbase, five frameshift mutations, which lead to premature stop codon and truncated product, three nonsense mutations, and a duplication, which causes a frameshift and premature termination. These mutations are disease causing, because they produce a nonfunctional, truncated protein. There is just one missense mutation, F98V substitution. The presence of a single, disease-causing missense mutation is not very informative, since this stretch is rather short and does not contain any hotspots for mutations, such as repetitive sequences or CpG dinucleotides. The loop is highly conserved.

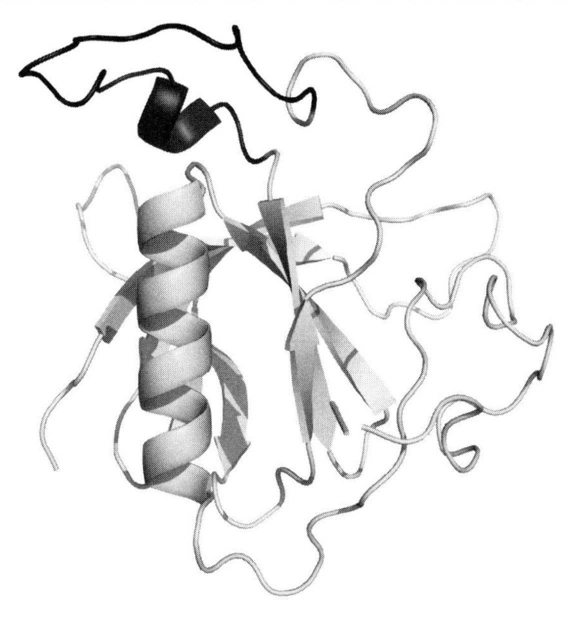

Figure 3.2. Amniotes specific insert in the PH domain of BTK. The insert is protruding up as a dark loop in the 1BTK structure containing the PH domain and BTK motif from human BTK. The loop is mainly nonstructured, but contains a short α-helix.

There are just minor variations in the sequence alignment, which suggests that selection pressure has maintained the sequence. Whether the underlying mechanism is functional or has a different origin is not clear.

There is also a missense SNP in the region, rs56035945 with R82K alteration, without known phenotype. Chicken, platypus and opossum Btks have a lysine at that site while there is arginine in the higher mammals. Arginine and lysine are both basic residues and frequently substituted by each other in protein families.

E. N-terminal regions of insect TFKs

D. melanogaster has three alternatively spliced mRNA forms, whereas a single form appears in other insects. The three mRNA variants code for two alternative protein products, because two of them differ just in their 5′-untranslated regions. The difference in the variants is that the longer form encompasses exons 1–5 and 9–16, while the shorter form contains exons 5–16. The longer protein variant aligns with that from *D. pseudoobscura*, while the shorter one with the proteins from *Anopheles gambiae* and *Culex pipiens quinquefasciatus*. Also other dipteran genomes might have alternatively spliced forms.

The N-terminus in insect proteins does not align at all with other TFK proteins before the beginning of the SH3 domain from whereon they align very well. Since this region is not conserved at all, and the whole N-terminus is truncated in the *Aedes aegypti* sequence, its specific function remains elusive.

V. SH3BP5—A CONSERVED NEGATIVE REGULATOR OF TFKs

The SH3-binding protein 5 interacts with Btk and MAP kinases. The ortholog sequences of human SH3BP5 protein were collected in a similar way as for TFKs. As there were closely related paralogs called SH3BP5L in several genomes they were included to the analysis.

In the tree, there are several protein groups (Fig. 3.3). An ortholog appears also in M. *brevicollis*, but does not fit to any of the groups. The insect and vertebrate groups for both SH3BP5 and SH3BP5L can be clearly distinguished along with a group in nematodes. The relationship of these big groups is more difficult to resolve. The simplest explanation for the tree is that there was a single SH3BP5 gene in the ancestors of eukaryotes and a duplication appeared very early. Both the duplicons were then preserved in insect as well as vertebrate lineages. The nematode group does not fit easily to this explanation. The role of both SH3BP5 and SH3BP5L proteins has to be significant, since they can be identified from all the important branches of eukaryotes. The detailed function of these proteins remains to be elucidated.

VI. THE ORIGIN OF PHOSPHOTYROSINE SIGNALING AND THE ROLE OF CYTOPLASMIC TYROSINE KINASES

The evolution of phosphotyrosine signaling suggests that more than 600 million years ago there was a common ancestor for the unicellular choanoflagellates and for multicellular metazoans, which had already developed this ability (King and Carroll, 2001; King *et al.*, 2008; Peterson and Butterfield, 2005; Pincus *et al.*, 2008). In some species, such as in yeast, tyrosine phosphorylation appears at a very low level, most likely due to promiscuity of serine/threonine kinases (Schieven *et al.*, 1986). In a recent report, Pincus *et al.* (2008) suggest that phosphatases and SH2 domains appeared first, whereas the enzymatic activity of tyrosine kinases developed later. The emergence of specific proteins resulted in the expansion of proteins and domains in cellular signaling. One third of all domains found in combination with SH2 domains in choanoflagellates are unique while 38% are shared with metazoans.

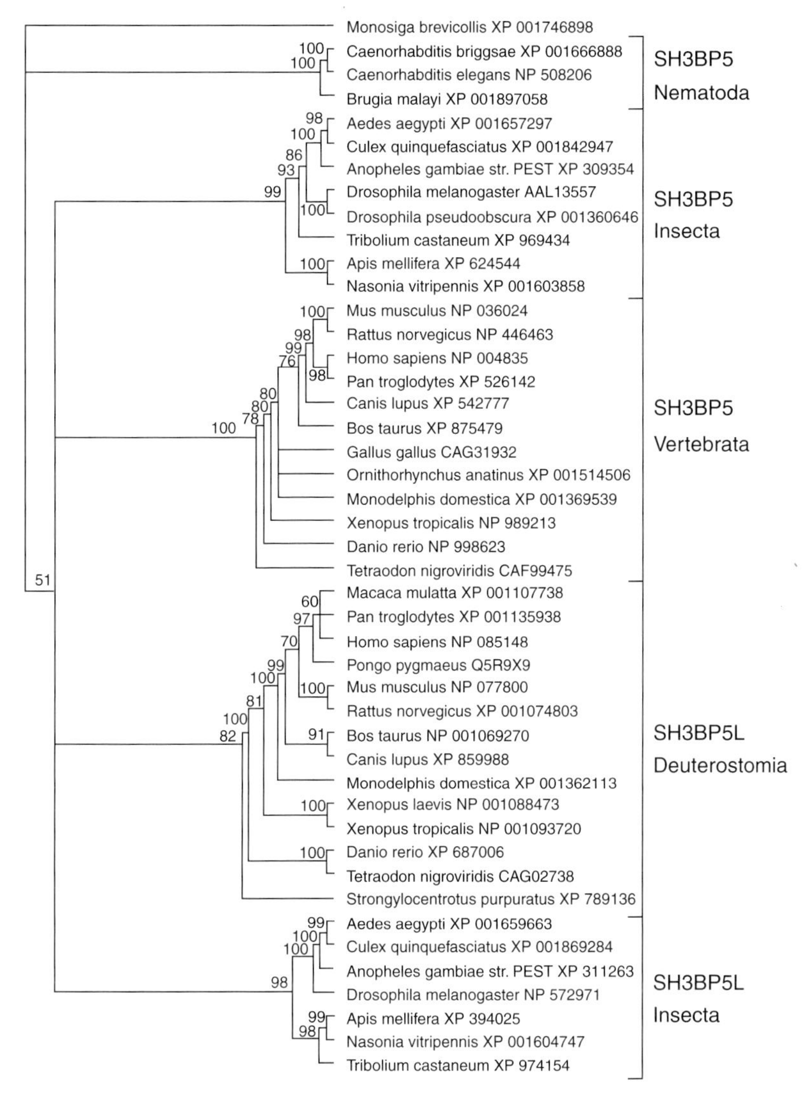

Figure 3.3. Phylogenetic relationship of the SH3-binding protein 5 and related sequences. Groups of SH3BP5 and SH3BP5L proteins from vertebrates and insects can be distinguished in addition to a distinct group of SH3BP5 proteins from nematodes. Further relationship of these groups is not clear. The analysis was carried out as described in Fig. 3.1.

Proteins active in tyrosine kinase-related signaling are quite abundant in choanoflagellates (King *et al.*, 2008). While most proteins in serine–threonine signaling pathways are common between metazoans and choanoflagellates, the opposite is true for tyrosine phosphorylation and several other intracellular signaling pathways, including many transcription factors.

Among choanoflagellates, SFKs and C-terminal Src kinase (Csk) were first reported in M. *ovata* (Segawa *et al.*, 2006). The M. *ovata* Src has transforming capability but is not negatively regulated by Csk. Biochemical characterization of the M. *brevicollis* Src and Csk indicated that the putative, regulatory C-terminal tyrosine is not phosphorylated (Li *et al.*, 2008).

Our report is the first to demonstrate the existence of a TFK in M. *brevicollis*, which has estimated to have 128 tyrosine kinase genes (King *et al.*, 2008). Although the role of the M. *brevicollis* TFK is unknown, its mere existence clearly suggests that it is functionally active in a unicellular organism. Since all domains of TFKs are conserved in this protein, it is possible that this kinase is already regulated by SFKs, PI3K, and PKC, like the metazoan counterparts. However, given the lack of Csk-induced control of SFKs in M. *brevicollis* as well as in M. *ovata*, it is equally possible that the regulation differs. Choanoflagellates also encode an SH3BP5-related molecule, which has not been functionally characterized. Thus, it is too early say whether this molecule suppresses the corresponding TFK. Functional studies will be needed to resolve this issue as well as the possibility that the choanoflagellate TFK can substitute for the loss of TFKs in metazoan cells.

Our study of TFKs reveals that these enzymes are ancient and their ancestor appeared already in choanoflagellates. TFK members are regulated by several proteins and they control numerous signaling pathways. More studies will be needed to investigate how the pathways in which TFKs currently participate originally obtained this property.

Acknowledgments

This work was supported by The Swedish Cancer Fund, The Wallenberg Foundation, the Swedish Science Council, the Stockholm County Council (research grant ALF-projektmedel), the European Union grant FP7-HEALTH-F2-2008-201549, the Medical Research Fund of Tampere University Hospital, and Academy of Finland.

References

Altman, A., Kaminski, S., Busuttil, V., Droin, N., Hu, J., Tadevosyan, Y., Hipskind, R. A., and Villalba, M. (2004). Positive feedback regulation of PLCγ1/Ca($^{2+}$) signaling by PKCθ in restimulated T cells via a Tec kinase-dependent pathway. *Eur. J. Immunol.* **34,** 2001–2011.

Altschul, S. F., Madden, T. L., Schaffer, A. A., Zhang, J., Zhang, Z., Miller, W., and Lipman, D. J. (1997). Gapped BLAST and PSI-BLAST: A new generation of protein database search programs. *Nucleic Acids Res.* **25,** 3389–3402.

Andreotti, A. H., Bunnell, S. C., Feng, S., Berg, L. J., and Schreiber, S. L. (1997). Regulatory intramolecular association in a tyrosine kinase of the Tec family. *Nature* **385,** 93–97.

Aparicio, S., Chapman, J., Stupka, E., Putnam, N., Chia, J. M., Dehal, P., Christoffels, A., Rash, S., Hoon, S., Smit, A., Gelpke, M. D., Roach, J., *et al.* (2002). Whole-genome shotgun assembly and analysis of the genome of *Fugu rubripes*. *Science* **297,** 1301–1310.

Atkinson, B. T., Ellmeier, W., and Watson, S. P. (2003). Tec regulates platelet activation by GPVI in the absence of Btk. *Blood* **102,** 3592–3599.

August, A., Sadra, A., Dupont, B., and Hanafusa, H. (1997). Src-induced activation of inducible T cell kinase (ITK) requires phosphatidylinositol 3-kinase activity and the pleckstrin homology domain of inducible T cell kinase. *Proc. Natl Acad. Sci. USA* **94,** 11227–11232.

Baba, K., Takeshita, A., Majima, K., Ueda, R., Kondo, S., Juni, N., and Yamamoto, D. (1999). The *Drosophila* Bruton's tyrosine kinase (Btk) homolog is required for adult survival and male genital formation. *Mol. Cell. Biol.* **19,** 4405–4413.

Bäckesjö, C. M., Vargas, L., Superti-Furga, G., and Smith, C. I. E. (2002). Phosphorylation of Bruton's tyrosine kinase by c-Abl. *Biochem. Biophys. Res. Commun.* **299,** 510–515.

Bajpai, U. D., Zhang, K., Teutsch, M., Sen, R., and Wortis, H. H. (2000). Bruton's tyrosine kinase links the B cell receptor to nuclear factor κB activation. *J. Exp. Med.* **191,** 1735–1744.

Baraldi, E., Djinovic Carugo, K., Hyvönen, M., Surdo, P. L., Riley, A. M., Potter, B. V., O'Brien, R., Ladbury, J. E., and Saraste, M. (1999). Structure of the PH domain from Bruton's tyrosine kinase in complex with inositol 1,3,4,5-tetrakisphosphate. *Structure* **7,** 449–460.

Behrsin, C. D., Bailey, M. L., Bateman, K. S., Hamilton, K. S., Wahl, L. M., Brandl, C. J., Shilton, B. H., and Litchfield, D. W. (2007). Functionally important residues in the peptidyl-prolyl isomerase Pin1 revealed by unigenic evolution. *J. Mol. Biol.* **365,** 1143–1162.

Berg, L. J., Finkelstein, L. D., Lucas, J. A., and Schwartzberg, P. L. (2005). Tec family kinases in T lymphocyte development and function. *Annu. Rev. Immunol.* **23,** 549–600.

Bögre, L., Okresz, L., Henriques, R., and Anthony, R. G. (2003). Growth signalling pathways in Arabidopsis and the AGC protein kinases. *Trends Plant Sci.* **8,** 424–431.

Bowes, J. B., Snyder, K. A., Segerdell, E., Gibb, R., Jarabek, C., Noumen, E., Pollet, N., and Vize, P. D. (2008). Xenbase: A Xenopus biology and genomics resource. *Nucleic Acids Res.* **36,** D761–D767.

Brazin, K. N., Fulton, D. B., and Andreotti, A. H. (2000). A specific intermolecular association between the regulatory domains of a Tec family kinase. *J. Mol. Biol.* **302,** 607–623.

Breheny, P. J., Laederach, A., Fulton, D. B., and Andreotti, A. H. (2003). Ligand specificity modulated by prolyl imide bond *cis/trans* isomerization in the Itk SH2 domain: A quantitative NMR study. *J. Am. Chem. Soc.* **125,** 15706–15707.

Broussard, C., Fleischacker, C., Horai, R., Chetana, M., Venegas, A. M., Sharp, L. L., Hedrick, S. M., Fowlkes, B. J., and Schwartzberg, P. L. (2006). Altered development of CD8+ T cell lineages in mice deficient for the Tec kinases Itk and Rlk. *Immunity* **25,** 93–104.

Brown, K., Long, J. M., Vial, S. C., Dedi, N., Dunster, N. J., Renwick, S. B., Tanner, A. J., Frantz, J. D., Fleming, M. A., and Cheetham, G. M. (2004). Crystal structures of interleukin-2 tyrosine kinase and their implications for the design of selective inhibitors. *J. Biol. Chem.* **279,** 18727–18732.

Brunner, C., Muller, B., and Wirth, T. (2005). Bruton's tyrosine kinase is involved in innate and adaptive immunity. *Histol. Histopathol.* **20,** 945–955.

Caenepeel, S., Charydczak, G., Sudarsanam, S., Hunter, T., and Manning, G. (2004). The mouse kinome: Discovery and comparative genomics of all mouse protein kinases. *Proc. Natl Acad. Sci. USA* **101,** 11707–11712.

Cetkovic, H., Muller, W. E., and Gamulin, V. (2004). Bruton tyrosine kinase-like protein, btksd, is present in the marine sponge *Suberites domuncula*. *Genomics* **83,** 743–745.

Chandrasekaran, V., and Beckendorf, S. K. (2005). Tec29 controls actin remodeling and endoreplication during invagination of the *Drosophila* embryonic salivary glands. *Development* **132,** 3515–3524.

Crosby, D., and Poole, A. W. (2002). Interaction of Bruton's tyrosine kinase and protein kinase Cθ in platelets. Cross-talk between tyrosine and serine/threonine kinases. *J. Biol. Chem.* **277,** 9958–9965.

Ekman, N., Lymboussaki, A., Västrik, I., Sarvas, K., Kaipainen, A., and Alitalo, K. (1997). Bmx tyrosine kinase is specifically expressed in the endocardium and the endothelium of large arteries. *Circulation* **96,** 1729–1732.

Ekman, N., Arighi, E., Rajantie, I., Saharinen, P., Riskimäki, A., Silvennoinen, O., and Alitalo, K. (2000). The Bmx tyrosine kinase is activated by IL-3 and G-CSF in a PI-3K dependent manner. *Oncogene* **19,** 4151–4158.

Ellmeier, W., Jung, S., Sunshine, M. J., Hatam, F., Xu, Y., Baltimore, D., Mano, H., and Littman, D. R. (2000). Severe B cell deficiency in mice lacking the tec kinase family members Tec and Btk. *J. Exp. Med.* **192,** 1611–1624.

Felices, M., and Berg, L. J. (2008). The Tec kinases Itk and Rlk regulate NKT cell maturation, cytokine production, and survival. *J. Immunol.* **180,** 3007–3018.

Felices, M., Falk, M., Kosaka, Y., and Berg, L. J. (2007). Tec kinases in T cell and mast cell signaling. *Adv. Immunol.* **93,** 145–184.

Finkelstein, L. D., and Schwartzberg, P. L. (2004). Tec kinases: Shaping T-cell activation through actin. *Trends Cell Biol.* **14,** 443–451.

Gibbs, R. A., Rogers, J., Katze, M. G., Bumgarner, R., Weinstock, G. M., Mardis, E. R., Remington, K. A., Strausberg, R. L., Venter, J. C., Wilson, R. K., Batzer, M. A., Bustamante, C. D., *et al.* (2007). Evolutionary and biomedical insights from the rhesus macaque genome. *Science* **316,** 222–234.

Gomez-Rodriguez, J., Readinger, J. A., Viorritto, I. C., Mueller, K. L., Houghtling, R. A., and Schwartzberg, P. L. (2007). Tec kinases, actin, and cell adhesion. *Immunol. Rev.* **218,** 45–64.

Guarnieri, D. J., Dodson, G. S., and Simon, M. A. (1998). SRC64 regulates the localization of a Tec-family kinase required for *Drosophila* ring canal growth. *Mol. Cell* **1,** 831–840.

Haire, R. N., Ohta, Y., Lewis, J. E., Fu, S. M., Kroisel, P., and Litman, G. W. (1994). TXK, a novel human tyrosine kinase expressed in T cells shares sequence identity with Tec family kinases and maps to 4p12. *Hum. Mol. Genet.* **3,** 897–901.

Haire, R. N., Strong, S. J., and Litman, G. W. (1997). Identification and characterization of a homologue of Bruton's tyrosine kinase, a Tec kinase involved in B-cell development, in a modern representative of a phylogenetically ancient vertebrate. *Immunogenetics* **46,** 349–351.

Haire, R. N., Strong, S. J., and Litman, G. W. (1998). Tec-family non-receptor tyrosine kinase expressed in zebrafish kidney. *Immunogenetics* **47,** 336–337.

Hamada, N., Bäckesjö, C. M., Smith, C. I. E., and Yamamoto, D. (2005). Functional replacement of *Drosophila* Btk29A with human Btk in male genital development and survival. *FEBS Lett.* **579,** 4131–4137.

Hamman, B. D., Pollok, B. A., Bennett, T., Allen, J., and Heim, R. (2002). Binding of a pleckstrin homology domain protein to phosphoinositide in membranes: A miniaturized FRET-based assay for drug screening. *J. Biomol. Screen.* **7,** 45–55.

Hanes, S. D., Shank, P. R., and Bostian, K. A. (1989). Sequence and mutational analysis of ESS1, a gene essential for growth in *Saccharomyces cerevisiae*. *Yeast* **5,** 55–72.

Hansson, H., Mattsson, P. T., Allard, P., Haapaniemi, P., Vihinen, M., Smith, C. I. E., and Härd, T. (1998). Solution structure of the SH3 domain from Bruton's tyrosine kinase. *Biochemistry* **37,** 2912–2924.

Hansson, H., Okoh, M. P., Smith, C. I. E., Vihinen, M., and Härd, T. (2001a). Intermolecular interactions between the SH3 domain and the proline-rich TH region of Bruton's tyrosine kinase. *FEBS Lett.* **489,** 67–70.

Hansson, H., Smith, C. I. E., and Härd, T. (2001b). Both proline-rich sequences in the TH region of Bruton's tyrosine kinase stabilize intermolecular interactions with the SH3 domain. *FEBS Lett.* **508,** 11–15.

Hao, S., and August, A. (2002). The proline rich region of the Tec homology domain of ITK regulates its activity. *FEBS Lett.* **525,** 53–58.

Hasan, M., Lopez-Herrera, G., Blomberg, K. E., Lindvall, J. M., Berglöf, A., Smith, C. I. E., and Vargas, L. (2008). Defective Toll-like receptor 9-mediated cytokine production in B cells from Bruton's tyrosine kinase-deficient mice. *Immunology* **123,** 239–249.

He, Y., Luo, Y., Tang, S., Rajantie, I., Salven, P., Heil, M., Zhang, R., Luo, D., Li, X., Chi, H., Yu, J., Carmeliet, P., *et al.* (2006). Critical function of Bmx/Etk in ischemia-mediated arteriogenesis and angiogenesis. *J. Clin. Invest.* **116,** 2344–2355.

Herman, P. K., Stack, J. H., and Emr, S. D. (1992). An essential role for a protein and lipid kinase complex in secretory protein sorting. *Trends Cell Biol.* **2,** 363–368.

Heyeck, S. D., and Berg, L. J. (1993). Developmental regulation of a murine T-cell-specific tyrosine kinase gene, *Tsk. Proc. Natl Acad. Sci. USA* **90,** 669–673.

Hirayama, T., Ohto, C., Mizoguchi, T., and Shinozaki, K. (1995). A gene encoding a phosphatidylinositol-specific phospholipase C is induced by dehydration and salt stress in *Arabidopsis thaliana. Proc. Natl Acad. Sci. USA* **92,** 3903–3907.

Holinski-Feder, E., Weiss, M., Brandau, O., Jedele, K. B., Nore, B., Bäckesjö, C. M., Vihinen, M., Hubbard, S. R., Belohradsky, B. H., Smith, C. I. E., and Meindl, A. (1998). Mutation screening of the *BTK* gene in 56 families with X-linked agammaglobulinemia (XLA): 47 unique mutations without correlation to clinical course. *Pediatrics* **101,** 276–284.

Hong, Z., and Verma, D. P. (1994). A phosphatidylinositol 3-kinase is induced during soybean nodule organogenesis and is associated with membrane proliferation. *Proc. Natl Acad. Sci. USA* **91,** 9617–9621.

Hu, Q., Davidson, D., Schwartzberg, P. L., Macchiarini, F., Lenardo, M. J., Bluestone, J. A., and Matis, L. A. (1995). Identification of Rlk, a novel protein tyrosine kinase with predominant expression in the T cell lineage. *J. Biol. Chem.* **270,** 1928–1934.

Huang, K. C., Cheng, H. T., Pai, M. T., Tzeng, S. R., and Cheng, J. W. (2006). Solution structure and phosphopeptide binding of the SH2 domain from the human Bruton's tyrosine kinase. *J. Biomol. NMR* **36,** 73–78.

Huang, Y. H., Grasis, J. A., Miller, A. T., Xu, R., Soonthornvacharin, S., Andreotti, A. H., Tsoukas, C. D., Cooke, M. P., and Sauer, K. (2007). Positive regulation of Itk PH domain function by soluble IP_4. *Science* **316,** 886–889.

Humphries, L. A., Dangelmaier, C., Sommer, K., Kipp, K., Kato, R. M., Griffith, N., Bakman, I., Turk, C. W., Daniel, J. L., and Rawlings, D. J. (2004). Tec kinases mediate sustained calcium influx via site-specific tyrosine phosphorylation of the phospholipase $C\gamma$ Src homology 2-Src homology 3 linker. *J. Biol. Chem.* **279,** 37651–37661.

Hyvönen, M., and Saraste, M. (1997). Structure of the PH domain and Btk motif from Bruton's tyrosine kinase: Molecular explanations for X-linked agammaglobulinaemia. *EMBO J.* **16,** 3396–3404.

Ingley, E. (2008). Src family kinases: Regulation of their activities, levels and identification of new pathways. *Biochim. Biophys. Acta* **1784,** 56–65.

Islam, T. C., and Smith, C. I. E. (2000). The cellular phenotype conditions Btk for cell survival or apoptosis signaling. *Immunol. Rev.* **178,** 49–63.

Jiang, X., Borgesi, R. A., McKnight, N. C., Kaur, R., Carpenter, C. L., and Balk, S. P. (2007). Activation of nonreceptor tyrosine kinase Bmx/Etk mediated by phosphoinositide 3-kinase, epidermal growth factor receptor, and erbb3 in prostate cancer cells. *J. Biol. Chem.* **282,** 32689–32698.

Jin, H., Webster, A. D., Vihinen, M., Sideras, P., Vorechovsky, I., Hammarström, L., Bernatowska-Matuszkiewicz, E., Smith, C. I. E., Bobrow, M., and Vetrie, D. (1995). Identification of Btk mutations in 20 unrelated patients with X-linked agammaglobulinaemia (XLA). *Hum. Mol. Genet.* **4,** 693–700.

Johannes, F. J., Hausser, A., Storz, P., Truckenmüller, L., Link, G., Kawakami, T., and Pfizenmaier, K. (1999). Bruton's tyrosine kinase (Btk) associates with protein kinase C μ. *FEBS Lett.* **461,** 68–72.

Jongstra-Bilen, J., Puig Cano, A., Hasija, M., Xiao, H., Smith, C. I. E., and Cybulsky, M. I. (2008). Dual functions of Bruton's tyrosine kinase and Tec kinase during Fcγ receptor-induced signaling and phagocytosis. *J. Immunol.* **181,** 288–298.

Joseph, R. E., Fulton, D. B., and Andreotti, A. H. (2007a). Mechanism and functional significance of Itk autophosphorylation. *J. Mol. Biol.* **373,** 1281–1292.

Joseph, R. E., Min, L., and Andreotti, A. H. (2007b). The linker between SH2 and kinase domains positively regulates catalysis of the Tec family kinases. *Biochemistry* **46,** 5455–5462.

Joseph, R. E., Min, L., Xu, R., Musselman, E. D., and Andreotti, A. H. (2007c). A remote substrate docking mechanism for the Tec family tyrosine kinases. *Biochemistry* **46,** 5595–5603.

Kane, L. P., and Watkins, S. C. (2005). Dynamic regulation of Tec kinase localization in membrane-proximal vesicles of a T cell clone revealed by total internal reflection fluorescence and confocal microscopy. *J. Biol. Chem.* **280,** 21949–21954.

Kang, S. W., Wahl, M. I., Chu, J., Kitaura, J., Kawakami, Y., Kato, R. M., Tabuchi, R., Tarakhovsky, A., Kawakami, T., Turck, C. W., Witte, O. N., and Rawlings, D. J. (2001). PKCβ modulates antigen receptor signaling via regulation of Btk membrane localization. *EMBO J.* **20,** 5692–5702.

Kawakami, Y., Yao, L., Han, W., and Kawakami, T. (1996). Tec family protein–tyrosine kinases and pleckstrin homology domains in mast cells. *Immunol. Lett.* **54,** 113–117.

Kawakami, Y., Kitaura, J., Hartman, S. E., Lowell, C. A., Siraganian, R. P., and Kawakami, T. (2000). Regulation of protein kinase CβI by two protein–tyrosine kinases, Btk and Syk. *Proc. Natl Acad. Sci. USA* **97,** 7423–7428.

Kim, Y. J., Sekiya, F., Poulin, B., Bae, Y. S., and Rhee, S. G. (2004). Mechanism of B-cell receptor-induced phosphorylation and activation of phospholipase C-γ2. *Mol. Cell. Biol.* **24,** 9986–9999.

King, N., and Carroll, S. B. (2001). A receptor tyrosine kinase from choanoflagellates: Molecular insights into early animal evolution. *Proc. Natl Acad. Sci. USA* **98,** 15032–15037.

King, N., Westbrook, M. J., Young, S. L., Kuo, A., Abedin, M., Chapman, J., Fairclough, S., Hellsten, U., Isogai, Y., Letunic, I., Marr, M., Pincus, D., *et al.* (2008). The genome of the choanoflagellate *Monosiga brevicollis* and the origin of metazoans. *Nature* **451,** 783–788.

Knapp, S., Mattsson, P. T., Christova, P., Berndt, K. D., Karshikoff, A., Vihinen, M., Smith, C. I. E., and Ladenstein, R. (1998). Thermal unfolding of small proteins with SH3 domain folding pattern. *Proteins* **31,** 309–319.

Kojima, T., Fukuda, M., Watanabe, Y., Hamazato, F., and Mikoshiba, K. (1997). Characterization of the pleckstrin homology domain of Btk as an inositol polyphosphate and phosphoinositide binding domain. *Biochem. Biophys. Res. Commun.* **236,** 333–339.

Korpi, M., Väliaho, J., and Vihinen, M. (2000). Structure-function effects in primary immunodeficiencies. *Scand. J. Immunol.* **52,** 226–232.

Koyanagi, M., Ono, K., Suga, H., Iwabe, N., and Miyata, T. (1998). Phospholipase C cDNAs from sponge and hydra: Antiquity of genes involved in the inositol phospholipid signaling pathway. *FEBS Lett.* **439,** 66–70.

Lachance, G., Levasseur, S., and Naccache, P. H. (2002). Chemotactic factor-induced recruitment and activation of Tec family kinases in human neutrophils. Implication of phosphatidylinositol 3-kinases. *J. Biol. Chem.* **277,** 21537–21541.

Laederach, A., Cradic, K. W., Brazin, K. N., Zamoon, J., Fulton, D. B., Huang, X. Y., and Andreotti, A. H. (2002). Competing modes of self-association in the regulatory domains of Bruton's tyrosine kinase: Intramolecular contact versus asymmetric homodimerization. *Protein Sci.* **11,** 36–45.

Laederach, A., Cradic, K. W., Fulton, D. B., and Andreotti, A. H. (2003). Determinants of intra versus intermolecular self-association within the regulatory domains of Rlk and Itk. *J. Mol. Biol.* **329,** 1011–1020.

Lappalainen, I., Thusberg, J., Shen, B., and Vihinen, M. (2008). Genome wide analysis of pathogenic SH2 domain mutations. *Proteins* **72,** 779–792.

Lee, S. H., Kim, T., Jeong, D., Kim, N., and Choi, Y. (2008). The Tec family tyrosine kinase Btk regulates RANKL-induced osteoclast maturation. *J. Biol. Chem.* **283,** 11526–11534.

Lehnes, K., Winder, A. D., Alfonso, C., Kasid, N., Simoneaux, M., Summe, H., Morgan, E., Iann, M. C., Duncan, J., Eagan, M., Tavaluc, R., Evans, C. H., Jr., *et al.* (2007). The effect of estradiol on *in vivo* tumorigenesis is modulated by the human epidermal growth factor receptor 2/phosphatidylinositol 3-kinase/Akt1 pathway. *Endocrinology* **148,** 1171–1180.

Leitges, M., Schmedt, C., Guinamard, R., Davoust, J., Schaal, S., Stabel, S., and Tarakhovsky, A. (1996). Immunodeficiency in protein kinase cβ—deficient mice. *Science* **273,** 788–791.

Levin, D. E., Fields, F. O., Kunisawa, R., Bishop, J. M., and Thorner, J. (1990). A candidate protein kinase C gene, PKC1, is required for the *S. cerevisiae* cell cycle. *Cell* **62,** 213–224.

Lewis, C. M., Broussard, C., Czar, M. J., and Schwartzberg, P. L. (2001). Tec kinases: Modulators of lymphocyte signaling and development. *Curr. Opin. Immunol.* **13,** 317–325.

Li, W., Young, S. L., King, N., and Miller, W. T. (2008). Signaling properties of a non-metazoan Src kinase and the evolutionary history of Src negative regulation. *J. Biol. Chem.* **283,** 15491–15501.

Liao, X. C., and Littman, D. R. (1995). Altered T cell receptor signaling and disrupted T cell development in mice lacking Itk. *Immunity* **3,** 757–769.

Lindvall, J. M., Blomberg, K. E., Väliaho, J., Vargas, L., Heinonen, J. E., Berglöf, A., Mohamed, A. J., Nore, B. F., Vihinen, M., and Smith, C. I. E. (2005). Bruton's tyrosine kinase: Cell biology, sequence conservation, mutation spectrum, sirna modifications, and expression profiling. *Immunol. Rev.* **203,** 200–215.

Lu, K. P., Hanes, S. D., and Hunter, T. (1996). A human peptidyl-prolyl isomerase essential for regulation of mitosis. *Nature* **380,** 544–547.

Lu, Y., Cuevas, B., Gibson, S., Khan, H., LaPushin, R., Imboden, J., and Mills, G. B. (1998). Phosphatidylinositol 3-kinase is required for CD28 but not CD3 regulation of the TEC family tyrosine kinase EMT/ITK/TSK: Functional and physical interaction of EMT with phosphatidylinositol 3-kinase. *J. Immunol.* **161,** 5404–5412.

Mallis, R. J., Brazin, K. N., Fulton, D. B., and Andreotti, A. H. (2002). Structural characterization of a proline-driven conformational switch within the Itk SH2 domain. *Nat. Struct. Biol.* **9,** 900–905.

Maniar, H. S., Vihinen, M., Webster, A. D., Nilsson, L., and Smith, C. I. E. (1995). Structural basis for X-linked agammaglobulinemia (XLA): Mutations at interacting Btk residues R562, W563, and A582. *Clin. Immunol. Immunopathol.* **76,** S198–S202.

Manna, D., Albanese, A., Park, W. S., and Cho, W. (2007). Mechanistic basis of differential cellular responses of phosphatidylinositol 3,4-bisphosphate- and phosphatidylinositol 3,4,5-trisphosphate-binding pleckstrin homology domains. *J. Biol. Chem.* **282,** 32093–32105.

Mano, H., Ishikawa, F., Nishida, J., Hirai, H., and Takaku, F. (1990). A novel protein–tyrosine kinase, tec, is preferentially expressed in liver. *Oncogene* **5,** 1781–1786.

Mano, H., Mano, K., Tang, B., Koehler, M., Yi, T., Gilbert, D. J., Jenkins, N. A., Copeland, N. G., and Ihle, J. N. (1993). Expression of a novel form of Tec kinase in hematopoietic cells and mapping of the gene to chromosome 5 near Kit. *Oncogene* **8,** 417–424.

Mansell, A., Smith, R., Doyle, S. L., Gray, P., Fenner, J. E., Crack, P. J., Nicholson, S. E., Hilton, D. J., O'Neill, L. A., and Hertzog, P. J. (2006). Suppressor of cytokine signaling 1 negatively regulates Toll-like receptor signaling by mediating Mal degradation. *Nat. Immunol.* **7,** 148–155.

Mao, C., Zhou, M., and Uckun, F. M. (2001). Crystal structure of Bruton's tyrosine kinase domain suggests a novel pathway for activation and provides insights into the molecular basis of X-linked agammaglobulinemia. *J. Biol. Chem.* **276,** 41435–41443.

Márquez, J. A., Smith, C. I. E., Petoukhov, M. V., Lo Surdo, P., Mattsson, P. T., Knekt, M., Westlund, A., Scheffzek, K., Saraste, M., and Svergun, D. I. (2003). Conformation of full-length Bruton tyrosine kinase (Btk) from synchrotron X-ray solution scattering. *EMBO J.* **22,** 4616–4624.

Matsushita, M., Yamadori, T., Kato, S., Takemoto, Y., Inazawa, J., Baba, Y., Hashimoto, S., Sekine, S., Arai, S., Kunikata, T., Kurimoto, M., Kishimoto, T., *et al.* (1998). Identification and characterization of a novel SH3-domain binding protein, Sab, which preferentially associates with Bruton's tyrosine kinase (BTK). *Biochem. Biophys. Res. Commun.* **245,** 337–343.

Mattsson, P. T., Vihinen, M., and Smith, C. I. E. (1996). X-linked agammaglobulinemia (XLA): A genetic tyrosine kinase (Btk) disease. *BioEssays* **18,** 825–834.

Mattsson, P. T., Lappalainen, I., Bäckesjö, C. M., Brockmann, E., Lauren, S., Vihinen, M., and Smith, C. I. E. (2000). Six X-linked agammaglobulinemia-causing missense mutations in the Src homology 2 domain of Bruton's tyrosine kinase: Phosphotyrosine-binding and circular dichroism analysis. *J. Immunol.* **164,** 4170–4177.

Melcher, M., Unger, B., Schmidt, U., Rajantie, I. A., Alitalo, K., and Ellmeier, W. (2008). Essential roles for the Tec family kinases Tec and Btk in M-CSF receptor signaling pathways that regulate macrophage survival. *J. Immunol.* **180,** 8048–8056.

Moarefi, I., LaFevre-Bernt, M., Sicheri, F., Huse, M., Lee, C. H., Kuriyan, J., and Miller, W. T. (1997). Activation of the Src-family tyrosine kinase Hck by SH3 domain displacement. *Nature* **385,** 650–653.

Nars, M., and Vihinen, M. (2001). Coevolution of the domains of cytoplasmic tyrosine kinases. *Mol. Biol. Evol.* **18,** 312–321.

Nore, B. F., Vargas, L., Mohamed, A. J., Brandén, L. J., Bäckesjö, C. M., Islam, T. C., Mattsson, P. T., Hultenby, K., Christensson, B., and Smith, C. I. E. (2000). Redistribution of Bruton's tyrosine kinase by activation of phosphatidylinositol 3-kinase and Rho-family GTPases. *Eur. J. Immunol.* **30,** 145–154.

Nore, B. F., Mattsson, P. T., Antonsson, P., Bäckesjö, C. M., Westlund, A., Lennartsson, J., Hansson, H., Low, P., Rönnstrand, L., and Smith, C. I. E. (2003). Identification of phosphorylation sites within the SH3 domains of Tec family tyrosine kinases. *Biochim. Biophys. Acta* **1645,** 123–132.

Notarangelo, L. D., Mella, P., Jones, A., de Saint Basile, G., Savoldi, G., Cranston, T., Vihinen, M., and Schumacher, R. F. (2001). Mutations in severe combined immune deficiency (SCID) due to JAK3 deficiency. *Hum. Mutat.* **18,** 255–263.

Oda, A., Ikeda, Y., Ochs, H. D., Druker, B. J., Ozaki, K., Handa, M., Ariga, T., Sakiyama, Y., Witte, O. N., and Wahl, M. I. (2000). Rapid tyrosine phosphorylation and activation of Bruton's tyrosine/Tec kinases in platelets induced by collagen binding or CD32 cross-linking. *Blood* **95,** 1663–1670.

Okoh, M. P., and Vihinen, M. (1999). Pleckstrin homology domains of tec family protein kinases. *Biochem. Biophys. Res. Commun.* **265,** 151–157.

Okoh, M. P., Kainulainen, L., Heiskanen, K., Isa, M. N., Varming, K., Ruuskanen, O., and Vihinen, M. (2002). Novel insertions of Bruton tyrosine kinase in patients with X-linked agammaglobulinemia. *Hum. Mutat.* **20,** 480–481.

Okoh, M. P., and Vihinen, M. (2002). Interaction between Btk TH and SH3 domain. *Biopolymers* **63,** 325–334.

Ortutay, C., Väliaho, J., Stenberg, K., and Vihinen, M. (2005). KinMutBase: A registry of disease-causing mutations in protein kinase domains. *Hum. Mutat.* **25,** 435–442.

Park, H., Wahl, M. I., Afar, D. E., Turck, C. W., Rawlings, D. J., Tam, C., Scharenberg, A. M., Kinet, J. P., and Witte, O. N. (1996). Regulation of Btk function by a major autophosphorylation site within the SH3 domain. *Immunity* **4,** 515–525.

Park, W. S., Heo, W. D., Whalen, J. H., O'Rourke, N. A., Bryan, H. M., Meyer, T., and Teruel, M. N. (2008). Comprehensive identification of PIP3-regulated PH domains from *C. elegans* to *H. sapiens* by model prediction and live imaging. *Mol. Cell* **30,** 381–392.

Pawson, T., and Scott, J. D. (2005). Protein phosphorylation in signaling-50 years and counting. *Trends Biochem. Sci.* **30,** 286–290.

Pawson, T., Gish, G. D., and Nash, P. (2001). SH2 domains, interaction modules and cellular wiring. *Trends Cell Biol.* **11,** 504–511.

Peterson, K. J., and Butterfield, N. J. (2005). Origin of the Eumetazoa: Testing ecological predictions of molecular clocks against the Proterozoic fossil record. *Proc. Natl Acad. Sci. USA* **102,** 9547–9552.

Petro, J. B., Rahman, S. M., Ballard, D. W., and Khan, W. N. (2000). Bruton's tyrosine kinase is required for activation of IκB kinase and nuclear factor κB in response to B cell receptor engagement. *J. Exp. Med.* **191,** 1745–1754.

Piirilä, H., Väliaho, J., and Vihinen, M. (2006). Immunodeficiency mutation databases (IDbases). *Hum. Mutat.* **27,** 1200–1208.

Pincus, D., Letunic, I., Bork, P., and Lim, W. A. (2008). Evolution of the phospho-tyrosine signaling machinery in premetazoan lineages. *Proc. Natl. Acad. Sci. USA* **105,** 9680–9684.

Pursglove, S. E., Mulhern, T. D., Mackay, J. P., Hinds, M. G., and Booker, G. W. (2002). The solution structure and intramolecular associations of the Tec kinase Src homology 3 domain. *J. Biol. Chem.* **277,** 755–762.

Putnam, N. H., Butts, T., Ferrier, D. E., Furlong, R. F., Hellsten, U., Kawashima, T., Robinson-Rechavi, M., Shoguchi, E., Terry, A., Yu, J. K., Benito-Gutierrez, E. L., Dubchak, I., *et al.* (2008). The amphioxus genome and the evolution of the chordate karyotype. *Nature* **453,** 1064–1071.

Qiu, Y., Robinson, D., Pretlow, T. G., and Kung, H. J. (1998). Etk/Bmx, a tyrosine kinase with a pleckstrin-homology domain, is an effector of phosphatidylinositol 3′-kinase and is involved in interleukin 6-induced neuroendocrine differentiation of prostate cancer cells. *Proc. Natl Acad. Sci. USA* **95,** 3644–3649.

Quintaje, S. B., and Orchard, S. (2008). The annotation of both human and mouse kinomes in UniProtKB/Swiss-Prot: One small step in manual annotation, one giant step for full comprehension of genomes. *Mol. Cell Proteomics* **7**(8), 1409–1419.

Rajantie, I., Ekman, N., Iljin, K., Arighi, E., Gunji, Y., Kaukonen, J., Palotie, A., Dewerchin, M., Carmeliet, P., and Alitalo, K. (2001). Bmx tyrosine kinase has a redundant function downstream of angiopoietin and vascular endothelial growth factor receptors in arterial endothelium. *Mol. Cell. Biol.* **21,** 4647–4655.

Rameh, L. E., Arvidsson, A., Carraway, K. L., III, Couvillon, A. D., Rathbun, G., Crompton, A., VanRenterghem, B., Czech, M. P., Ravichandran, K. S., Burakoff, S. J., Wang, D. S., Chen, C. S., *et al.* (1997). A comparative analysis of the phosphoinositide binding specificity of pleckstrin homology domains. *J. Biol. Chem.* **272,** 22059–22066.

Rawlings, D. J., Saffran, D. C., Tsukada, S., Largaespada, D. A., Grimaldi, J. C., Cohen, L., Mohr, R. N., Bazan, J. F., Howard, M., Copeland, N. G., *et al.* (1993). Mutation of unique region of Bruton's tyrosine kinase in immunodeficient XID mice. *Science* **261,** 358–361.

Robinson, D., He, F., Pretlow, T., and Kung, H. J. (1996). A tyrosine kinase profile of prostate carcinoma. *Proc. Natl Acad. Sci. USA* **93,** 5958–5962.

Roulier, E. M., Panzer, S., and Beckendorf, S. K. (1998). The Tec29 tyrosine kinase is required during *Drosophila* embryogenesis and interacts with Src64 in ring canal development. *Mol. Cell* **1,** 819–829.

Sable, C. L., Filippa, N., Filloux, C., Hemmings, B. A., and Van Obberghen, E. (1998). Involvement of the pleckstrin homology domain in the insulin-stimulated activation of protein kinase B. *J. Biol. Chem.* **273,** 29600–29606.

Saharinen, P., Ekman, N., Sarvas, K., Parker, P., Alitalo, K., and Silvennoinen, O. (1997). The Bmx tyrosine kinase induces activation of the Stat signaling pathway, which is specifically inhibited by protein kinase Cδ. *Blood* **90,** 4341–4353.

Saito, K., Scharenberg, A. M., and Kinet, J. P. (2001). Interaction between the Btk PH domain and phosphatidylinositol-3,4,5-trisphosphate directly regulates Btk. *J. Biol. Chem.* **276,** 16201–16206.

Salim, K., Bottomley, M. J., Querfurth, E., Zvelebil, M. J., Gout, I., Scaife, R., Margolis, R. L., Gigg, R., Smith, C. I. E., Driscoll, P. C., Waterfield, M. D., and Panayotou, G. (1996). Distinct specificity in the recognition of phosphoinositides by the pleckstrin homology domains of dynamin and Bruton's tyrosine kinase. *EMBO J.* **15,** 6241–6250.

Schaeffer, E. M., Debnath, J., Yap, G., McVicar, D., Liao, X. C., Littman, D. R., Sher, A., Varmus, H. E., Lenardo, M. J., and Schwartzberg, P. L. (1999). Requirement for Tec kinases Rlk and Itk in T cell receptor signaling and immunity. *Science* **284,** 638–641.

Schieven, G., Thorner, J., and Martin, G. S. (1986). Protein–tyrosine kinase activity in *Saccharomyces cerevisiae*. *Science* **231,** 390–393.

Schmidt, U., van den Akker, E., Parren-van Amelsvoort, M., Litos, G., de Bruijn, M., Gutierrez, L., Hendriks, R. W., Ellmeier, W., Lowenberg, B., Beug, H., and von Lindern, M. (2004a). Btk is required for an efficient response to erythropoietin and for SCF-controlled protection against TRAIL in erythroid progenitors. *J. Exp. Med.* **199,** 785–795.

Schmidt, U., Boucheron, N., Unger, B., and Ellmeier, W. (2004b). The role of Tec family kinases in myeloid cells. *Int. Arch. Allergy Immunol.* **134,** 65–78.

Seet, B. T., Dikic, I., Zhou, M. M., and Pawson, T. (2006). Reading protein modifications with interaction domains. *Nat. Rev. Mol. Cell Biol.* **7,** 473–483.

Segawa, Y., Suga, H., Iwabe, N., Oneyama, C., Akagi, T., Miyata, T., and Okada, M. (2006). Functional development of Src tyrosine kinases during evolution from a unicellular ancestor to multicellular animals. *Proc. Natl Acad. Sci. USA* **103,** 12021–12026.

Serfas, M. S., and Tyner, A. L. (2003). Brk, Srm, Frk, and Src42A form a distinct family of intracellular Src-like tyrosine kinases. *Oncol. Res.* **13,** 409–419.

Severin, A., Fulton, D. B., and Andreotti, A. H. (2008). Murine Itk SH3 domain. *J. Biomol. NMR* **40,** 285–290.

Shinohara, M., Koga, T., Okamoto, K., Sakaguchi, S., Arai, K., Yasuda, H., Takai, T., Kodama, T., Morio, T., Geha, R. S., Kitamura, D., Kurosaki, T., *et al.* (2008). Tyrosine kinases Btk and Tec regulate osteoclast differentiation by linking RANK and ITAM signals. *Cell* **132,** 794–806.

Sicheri, F., Moarefi, I., and Kuriyan, J. (1997). Crystal structure of the Src family tyrosine kinase Hck. *Nature* **385,** 602–609.

Siliciano, J. D., Morrow, T. A., and Desiderio, S. V. (1992). ITK, a T-cell-specific tyrosine kinase gene inducible by interleukin 2. *Proc. Natl Acad. Sci. USA* **89,** 11194–11198.

Sinka, R., Jankovics, F., Somogyi, K., Szlanka, T., Lukacsovich, T., and Erdelyi, M. (2002). poirot, a new regulatory gene of *Drosophila* oskar acts at the level of the short Oskar protein isoform. *Development* **129,** 3469–3478.

Smith, C. I. E., Baskin, B., Humire-Greiff, P., Zhou, J. N., Olsson, P. G., Maniar, H. S., Kjellen, P., Lambris, J. D., Christensson, B., Hammarström, L., *et al.* (1994a). Expression of Bruton's agammaglobulinemia tyrosine kinase gene, BTK, is selectively down-regulated in T lymphocytes and plasma cells. *J. Immunol.* **152,** 557–565.

Smith, C. I. E., Islam, K. B., Vorechovsky, I., Olerup, O., Wallin, E., Rabbani, H., Baskin, B., and Hammarström, L. (1994b). X-linked agammaglobulinemia and other immunoglobulin deficiencies. *Immunol. Rev.* **138,** 159–183.

Smith, C. I. E., Islam, T. C., Mattsson, P. T., Mohamed, A. J., Nore, B. F., and Vihinen, M. (2001). The Tec family of cytoplasmic tyrosine kinases: Mammalian Btk, Bmx, Itk, Tec, Txk and homologs in other species. *BioEssays* **23,** 436–446.

Speletas, M., Kanariou, M., Kanakoudi-Tsakalidou, F., Papadopoulou-Alataki, E., Arvanitidis, K., Pardali, E., Constantopoulos, A., Kartalis, G., Vihinen, M., Sideras, P., and Ritis, K. (2001). Analysis of Btk mutations in patients with X-linked agammaglobulinaemia (XLA) and determination of carrier status in normal female relatives: A nationwide study of Btk deficiency in Greece. *Scand. J. Immunol.* **54,** 321–327.

Sprague, J., Bayraktaroglu, L., Bradford, Y., Conlin, T., Dunn, N., Fashena, D., Frazer, K., Haendel, M., Howe, D. G., Knight, J., Mani, P., Moxon, S. A., *et al.* (2008). The Zebrafish Information Network: The zebrafish model organism database provides expanded support for genotypes and phenotypes. *Nucleic Acids Res.* **36,** D768–D772.

Stephens, L. R., Jackson, T. R., and Hawkins, P. T. (1993). Agonist-stimulated synthesis of phosphatidylinositol(3,4,5)-trisphosphate: A new intracellular signalling system? *Biochim. Biophys. Acta* **1179,** 27–75.

Stoica, G. E., Franke, T. F., Moroni, M., Mueller, S., Morgan, E., Iann, M. C., Winder, A. D., Reiter, R., Wellstein, A., Martin, M. B., and Stoica, A. (2003). Effect of estradiol on estrogen receptor-alpha gene expression and activity can be modulated by the ErbB2/PI 3-K/Akt pathway. *Oncogene* **22,** 7998–8011.

Strausberg, R. L., Feingold, E. A., Grouse, L. H., Derge, J. G., Klausner, R. D., Collins, F. S., Wagner, L., Shenmen, C. M., Schuler, G. D., Altschul, S. F., Zeeberg, B., Buetow, K. H., *et al.* (2002). Generation and initial analysis of more than 15,000 full-length human and mouse cDNA sequences. *Proc. Natl. Acad. Sci. USA* **99,** 16899–16903.

Swofford, D. (2003). "PAUP*. Phylogenetic Analysis Using Parsimony (* and Other Methods). Version 4." Sinauer Associates, Sunderland, MA.

Tamagnone, L., Lahtinen, I., Mustonen, T., Virtaneva, K., Francis, F., Muscatelli, F., Alitalo, R., Smith, C. I. E., Larsson, C., and Alitalo, K. (1994). BMX, a novel nonreceptor tyrosine kinase gene of the BTK/ITK/TEC/TXK family located in chromosome Xp22.2. *Oncogene* **9,** 3683–3688.

Tasma, I. M., Brendel, V., Whitham, S. A., and Bhattacharyya, M. K. (2008). Expression and evolution of the phosphoinositide-specific phospholipase C gene family in *Arabidopsis thaliana*. *Plant Physiol. Biochem.* **46,** 627–637.

Thomas, J. H., and Wieschaus, E. (2004). src64 and tec29 are required for microfilament contraction during *Drosophila* cellularization. *Development* **131,** 863–871.

Thomas, J. D., Sideras, P., Smith, C. I. E., Vorechovsky, I., Chapman, V., and Paul, W. E. (1993). Colocalization of X-linked agammaglobulinemia and X-linked immunodeficiency genes. *Science* **261,** 355–358.

Tomlinson, M. G., Heath, V. L., Turck, C. W., Watson, S. P., and Weiss, A. (2004). SHIP family inositol phosphatases interact with and negatively regulate the Tec tyrosine kinase. *J. Biol. Chem.* **279,** 55089–55096.

Tsukada, S., Saffran, D. C., Rawlings, D. J., Parolini, O., Allen, R. C., Klisak, I., Sparkes, R. S., Kubagawa, H., Mohandas, T., Quan, S., *et al.* (1993). Deficient expression of a B cell cytoplasmic tyrosine kinase in human X-linked agammaglobulinemia. *Cell* **72,** 279–290.

Uckun, F. M. (1998). Bruton's tyrosine kinase (BTK) as a dual-function regulator of apoptosis. *Biochem. Pharmacol.* **56,** 683–691.

Väliaho, J., Smith, C. I. E., and Vihinen, M. (2006). BTKbase: The mutation database for X-linked agammaglobulinemia. *Hum. Mutat.* **27,** 1209–1217.

Vihinen, M., and Durandy, A. (2006). Primary Immunodeficiencies: Genotype-Phenotype Correlations. *In* "Immunogenomics and Human Disease" (A. Falus, ed.). John Wiley & Sons Inc., Hoboken, New Jersey. Vol. 1, pp. 443–460.

Vihinen, M., Villa, A., Mella, P., Schumacher, R. F., Savoldi, G., O'Shea, J. J., Candotti, F., and Notarangelo, L. D. (2000). Molecular modeling of the Jak3 kinase domains and structural basis for severe combined immunodeficiency. *Clin. Immunol.* **96,** 108–118.

Vorechovsky, I., Vihinen, M., de Saint Basile, G., Honsova, S., Hammarstrom, L., Muller, S., Nilsson, L., Fischer, A., and Smith, C. I. (1995). DNA-based mutation analysis of Bruton's tyrosine kinase gene in patients with X-linked agammaglobulinaemia. *Hum. Mol. Genet.* **4,** 51–58.

van Dijk, T. B., van Den Akker, E., Amelsvoort, M. P., Mano, H., Lowenberg, B., and von Lindern, M. (2000). Stem cell factor induces phosphatidylinositol 3′-kinase-dependent Lyn/Tec/Dok-1 complex formation in hematopoietic cells. *Blood* **96,** 3406–3413.

Vanhaesebroeck, B., Leevers, S. J., Panayotou, G., and Waterfield, M. D. (1997). Phosphoinositide 3-kinases: A conserved family of signal transducers. *Trends Biochem. Sci.* **22,** 267–272.

Varnai, P., Bondeva, T., Tamas, P., Toth, B., Buday, L., Hunyady, L., and Balla, T. (2005). Selective cellular effects of overexpressed pleckstrin-homology domains that recognize Ptdins(3,4,5)P_3 suggest their interaction with protein binding partners. *J. Cell Sci.* **118,** 4879–4888.

Venkataraman, C., Chen, X. C., Na, S., Lee, L., Neote, K., and Tan, S. L. (2006). Selective role of PKCβ enzymatic function in regulating cell survival mediated by B cell antigen receptor cross-linking. *Immunol. Lett.* **105,** 83–89.

Vetrie, D., Vorechovsky, I., Sideras, P., Holland, J., Davies, A., Flinter, F., Hammarström, L., Kinnon, C., Levinsky, R., Bobrow, M., *et al.* (1993). The gene involved in X-linked agammaglobulinaemia is a member of the src family of protein–tyrosine kinases. *Nature* **361,** 226–233.

Vihinen, M., Nilsson, L., and Smith, C. I. E. (1994a). Tec homology (TH) adjacent to the PH domain. *FEBS Lett.* **350,** 263–265.

Vihinen, M., Nilsson, L., and Smith, C. I. E. (1994b). Structural basis of SH2 domain mutations in X-linked agammaglobulinemia. *Biochem. Biophys. Res. Commun.* **205,** 1270–1277.

Vihinen, M., Vetrie, D., Maniar, H. S., Ochs, H. D., Zhu, Q., Vorechovsky, I., Webster, A. D., Notarangelo, L. D., Nilsson, L., Sowadski, J. M., *et al.* (1994c). Structural basis for chromosome X-linked agammaglobulinemia: A tyrosine kinase disease. *Proc. Natl Acad. Sci. USA* **91,** 12803–12807.

Vihinen, M., Cooper, M. D., de Saint Basile, G., Fischer, A., Good, R. A., Hendriks, R. W., Kinnon, C., Kwan, S. P., Litman, G. W., Notarangelo, L. D., *et al.* (1995a). BTKbase: A database of XLA-causing mutations. *Immunol. Today* **16,** 460–465.

Vihinen, M., Zvelebil, M. J., Zhu, Q., Brooimans, R. A., Ochs, H. D., Zegers, B. J., Nilsson, L., Waterfield, M. D., and Smith, C. I. E. (1995b). Structural basis for pleckstrin homology domain mutations in X-linked agammaglobulinemia. *Biochemistry* **34,** 1475–1481.

Vihinen, M., Iwata, T., Kinnon, C., Kwan, S. P., Ochs, H. D., Vorechovsky, I., and Smith, C. I. E. (1996). BTKbase, mutation database for X-linked agammaglobulinemia (XLA). *Nucleic Acids Res.* **24,** 160–165.

Vihinen, M., Nore, B. F., Mattsson, P. T., Bäckesjö, C. M., Nars, M., Koutaniemi, S., Watanabe, C., Lester, T., Jones, A., Ochs, H. D., and Smith, C. I. E. (1997a). Missense mutations affecting a conserved cysteine pair in the TH domain of Btk. *FEBS Lett.* **413,** 205–210.

Vihinen, M., Belohradsky, B. H., Haire, R. N., Holinski-Feder, E., Kwan, S. P., Lappalainen, I., Lehväslaiho, H., Lester, T., Meindl, A., Ochs, H. D., Ollila, J., Vorechovsky, I., *et al.* (1997b). BTKbase, mutation database for X-linked agammaglobulinemia (XLA). *Nucleic Acids Res.* **25,** 166–171.

Vihinen, M., Brandau, O., Brandén, L. J., Kwan, S. P., Lappalainen, I., Lester, T., Noordzij, J. G., Ochs, H. D., Ollila, J., Pienaar, S. M., Riikonen, P., Saha, B. K., *et al.* (1998). BTKbase, mutation database for X-linked agammaglobulinemia (XLA). *Nucleic Acids Res.* **26,** 242–247.

Vihinen, M., Kwan, S. P., Lester, T., Ochs, H. D., Resnick, I., Väliaho, J., Conley, M. E., and Smith, C. I. E. (1999). Mutations of the human *BTK* gene coding for Bruton tyrosine kinase in X-linked agammaglobulinemia. *Hum. Mutat.* **13,** 280–285.

Vihinen, M., Arredondo-Vega, F. X., Casanova, J. L., Etzioni, A., Giliani, S., Hammarström, L., Hershfield, M. S., Heyworth, P. G., Hsu, A. P., Lähdesmäki, A., Lappalainen, I., Notarangelo, L. D., *et al.* (2001). Primary immunodeficiency mutation databases. *Adv. Genet.* **43,** 103–188.

Vorechovsky, I., Luo, L., Hertz, J. M., Frøland, S. S., Klemola, T., Fiorini, M., Quinti, I., Paganelli, R., Ozsahin, H., Hammarström, L., Webster, A. D., and Smith, C. I. E. (1997). Mutation pattern in the Bruton's tyrosine kinase gene in 26 unrelated patients with X-linked agammaglobulinemia. *Hum. Mutat.* **9,** 418–425.

Wang, Q., Deloia, M. A., Kang, Y., Litchke, C., Zhang, N., Titus, M. A., and Walters, K. J. (2007). The SH3 domain of a M7 interacts with its C-terminal proline-rich region. *Protein Sci.* **16,** 189–196.

Watanabe, N., Nakajima, H., Suzuki, H., Oda, A., Matsubara, Y., Moroi, M., Terauchi, Y., Kadowaki, T., Koyasu, S., Ikeda, Y., and Handa, M. (2003). Functional phenotype of phosphoinositide 3-kinase p85α-null platelets characterized by an impaired response to GP VI stimulation. *Blood* **102,** 541–548.

Welters, P., Takegawa, K., Emr, S. D., and Chrispeels, M. J. (1994). AtVPS34, a phosphatidylinositol 3-kinase of *Arabidopsis thaliana,* is an essential protein with homology to a calcium-dependent lipid binding domain. *Proc. Natl Acad. Sci. USA* **91,** 11398–11402.

Wilcox, H. M., and Berg, L. J. (2003). Itk phosphorylation sites are required for functional activity in primary T cells. *J. Biol. Chem.* **278,** 37112–37121.

Williams, J. G., and Zvelebil, M. (2004). SH2 domains in plants imply new signalling scenarios. *Trends Plant Sci.* **9,** 161–163.

Williams, J. C., Weijland, A., Gonfloni, S., Thompson, A., Courtneidge, S. A., Superti-Furga, G., and Wierenga, R. K. (1997). The 2.35 Å crystal structure of the inactivated form of chicken Src: A dynamic molecule with multiple regulatory interactions. *J. Mol. Biol.* **274,** 757–775.

Wiltshire, C., Matsushita, M., Tsukada, S., Gillespie, D. A., and May, G. H. (2002). A new c-Jun N-terminal kinase (JNK)-interacting protein, Sab (SH3BP5), associates with mitochondria. *Biochem. J.* **367,** 577–585.

Wiltshire, C., Gillespie, D. A., and May, G. H. (2004). Sab (SH3BP5), a novel mitochondria-localized JNK-interacting protein. *Biochem. Soc. Trans.* **32,** 1075–1077.

Xu, W., Harrison, S. C., and Eck, M. J. (1997). Three-dimensional structure of the tyrosine kinase c-Src. *Nature* **385,** 595–602.

Xue, L. Y., Qiu, Y., He, J., Kung, H. J., and Oleinick, N. L. (1999). Etk/Bmx, a PH-domain containing tyrosine kinase, protects prostate cancer cells from apoptosis induced by photodynamic therapy or thapsigargin. *Oncogene* **18,** 3391–3398.

Yamada, N., Kawakami, Y., Kimura, H., Fukamachi, H., Baier, G., Altman, A., Kato, T., Inagaki, Y., and Kawakami, T. (1993). Structure and expression of novel protein–tyrosine kinases, Emb and Emt, in hematopoietic cells. *Biochem. Biophys. Res. Commun.* **192,** 231–240.

Yamadori, T., Baba, Y., Matsushita, M., Hashimoto, S., Kurosaki, M., Kurosaki, T., Kishimoto, T., and Tsukada, S. (1999). Bruton's tyrosine kinase activity is negatively regulated by Sab, the Btk-SH3 domain-binding protein. *Proc. Natl. Acad. Sci. USA* **96,** 6341–6346.

Yang, X. L., Zhang, Y. L., Lai, Z. S., Xing, F. Y., and Liu, Y. H. (2003). Pleckstrin homology domain of G protein-coupled receptor kinase-2 binds to PKC and affects the activity of PKC kinase. *World J. Gastroenterol.* **9,** 800–803.

Yao, L., Kawakami, Y., and Kawakami, T. (1994). The pleckstrin homology domain of Bruton tyrosine kinase interacts with protein kinase C. *Proc. Natl Acad. Sci. USA* **91,** 9175–9179.

Yao, J. L., Kops, O., Lu, P. J., and Lu, K. P. (2001). Functional conservation of phosphorylation-specific prolyl isomerases in plants. *J. Biol. Chem.* **276,** 13517–13523.

Yu, L., Mohamed, A. J., Vargas, L., Berglöf, A., Finn, G., Lu, K. P., and Smith, C. I. E. (2006). Regulation of Bruton tyrosine kinase by the peptidylprolyl isomerase Pin1. *J. Biol. Chem.* **281,** 18201–18207.

Yu, L., Mohamed, A. J., Simonson, O. E., Vargas, L., Blomberg, K. E., Bjorkstrand, B., Arteaga, H. J., Nore, B. F., and Smith, C. I. E. (2008). Proteasome-dependent autoregulation of Bruton tyrosine kinase (Btk) promoter via NF-κB. *Blood* **111,** 4617–4626.

Zegzouti, H., Li, W., Lorenz, T. C., Xie, M., Payne, C. T., Smith, K., Glenny, S., Payne, G. S., and Christensen, S. K. (2006). Structural and functional insights into the regulation of *Arabidopsis* AGC VIIIa kinases. *J. Biol. Chem.* **281,** 35520–35530.

Zhu, Q., Zhang, M., Rawlings, D. J., Vihinen, M., Hagemann, T., Saffran, D. C., Kwan, S. P., Nilsson, L., Smith, C. I. E., Witte, O. N., Chen, S. H., and Ochs, H. D. (1994). Deletion within the Src homology domain 3 of Bruton's tyrosine kinase resulting in X-linked agammaglobulinemia (XLA). *J. Exp. Med.* **180,** 461–470.

4

Therapeutic Approaches to Ion Channel Diseases

Diana Conte Camerino, Jean-François Desaphy, Domenico Tricarico, Sabata Pierno, and Antonella Liantonio

Division of Pharmacology, Department of Pharmacobiology, Faculty of Pharmacy, University of Bari, I-70125 Bari, Italy

Advances in Genetics, Vol. 64

0065-2660/08 $35.00
DOI: 10.1016/S0065-2660(08)00804-3

ABSTRACT

More than 400 genes are known that encode ion channel subunits. In addition, alternative splicing and heteromeric assembly of different subunits increase tremendously the variety of ion channels. Such many channels are needed to accomplish very complex cellular functions, whereas dysfunction of ion channels are key events in many pathological processes. The recent discovery of ion channelopathies, which, in its more stringent definition, encloses monogenic disorders due to mutations in ion channel genes, has largely contributed to our understanding of the function of the various channel subtypes and of the role of ion channels in multigenic or acquired diseases. Last but not least, ion channels are the main targets of many drugs already used in the clinics. Most of these drugs were introduced in therapy based on the experience acquired quite empirically, and many were discovered afterward to target ion channels. Now, intense research is being conducted to develop new drugs acting selectively on ion channel subtypes and aimed at the understanding of the intimate drug–channel interaction. In this review, we first focus on the pharmacotherapy of ion channel diseases, which includes many drugs targeting ion channels. Then, we describe the molecular pharmacology of ion channels, including the more recent advancement in drug development. Among the newest aspect of ion channel pharmacology, we draw attention to how polymorphisms or mutations in ion channel genes may modify sensitivity to drugs, opening the way toward the development of pharmacogenetics.

I. INTRODUCTION

More than 400 genes are known that encode ion channel subunits. In addition, alternative splicing and heteromeric assembly of different subunits increase tremendously the variety of ion channels. Such many channels are needed to accomplish very complex cellular functions, which include the control of cell electrical excitability, hormone and solute secretion, cell volume, cell proliferation and death, and many others. Dysfunction of ion channels are key events in many pathological processes. The recent discovery of ion channelopathies, which, in its more stringent definition, enclose monogenic disorders due to mutations in ion channel genes, has largely contributed to our understanding of the function of various channel subtypes and of the role of ion channels in multigenic or acquired diseases. Last but not least, ion channels are the main targets of many drugs already used in the clinics. Nonetheless, it is widely acknowledged that ion channels constitute an underexploited therapeutic field and that the number of drugs targeting ion channels is likely to increase.

Indeed, ion channel function is modulated by many natural products of the animal and vegetal kingdoms, which contribute to the dangerous effects of poisons or the beneficial effects of medicinal herbs. Once isolated, these lead compounds have served to the synthesis of more specific ligands with fewer side effects. Many of these drugs have been used empirically and their effects on ion channels have been revealed only afterward. More recently, high-throughput technologies have been used to identify new ligands for specific ion channels identified as being critically involved in diseases.

The aim of this review is twofold. First, we focus on the pharmacotherapy of ion channel diseases that includes many drugs targeting ion channels. Second, we describe the molecular pharmacology of ion channels including the more recent advancement in drug development.

II. PHARMACOTHERAPY OF ION CHANNELOPATHIES

A. Muscle channelopathies

Ion channels play a critical role in setting and maintaining the resting membrane potential and controlling excitability of skeletal muscles (Fig. 4.1). Muscle channelopathies are thus characterized by hyperexcitability (nondystrophic myotonia) or nonexcitability (periodic paralyses). Alteration of excitability due to ion channel dysfunction secondary to genetic alteration in other genes is also observed in the myotonic dystrophies. In addition, malignant hyperthermia (MH) and central core disease (CCD) are distinct muscle channelopathies associated with an altered intracellular excitation–contraction coupling mechanisms. Currently, the drug therapy is based on the use of drugs controlling the myotonia or the attack of weakness in periodic paralysis and in preventing the attacks in the MH and CCD diseases. The drugs of skeletal muscle channelopathies are used as off-label or as orphan drugs. The first group of drugs, largely used in the peripheral and CNS channelopathies, are represented by acetazolamide (ACTZ) and dichlorphenamide (DCP), well-known carbonic anhydrase enzyme inhibitors, which also target K^+ channels (KChs). The preferred antimyotonic drugs are Na^+ channel (NaCh) blockers, such as mexiletine and flecainide, while others such as taurine may enhance the Cl^- conductance. The third class of drugs are K^+-loss diuretics, such as hydrochlorothiazide and related analogues, or K^+-sparing diuretics, such as triamterene and spironolactone, which mostly act by correcting the abnormal serum K^+ level. The β_2-adrenoceptor agonists, such as salbutamol and terbutaline, that act by repolarization/relaxation of skeletal muscle fibers, represent additional treatment in hyperkalemic periodic paralysis. Inhibitors of the intracellular Ca^{2+} release channels such as dantrolene are used to treat MH and CCD diseases.

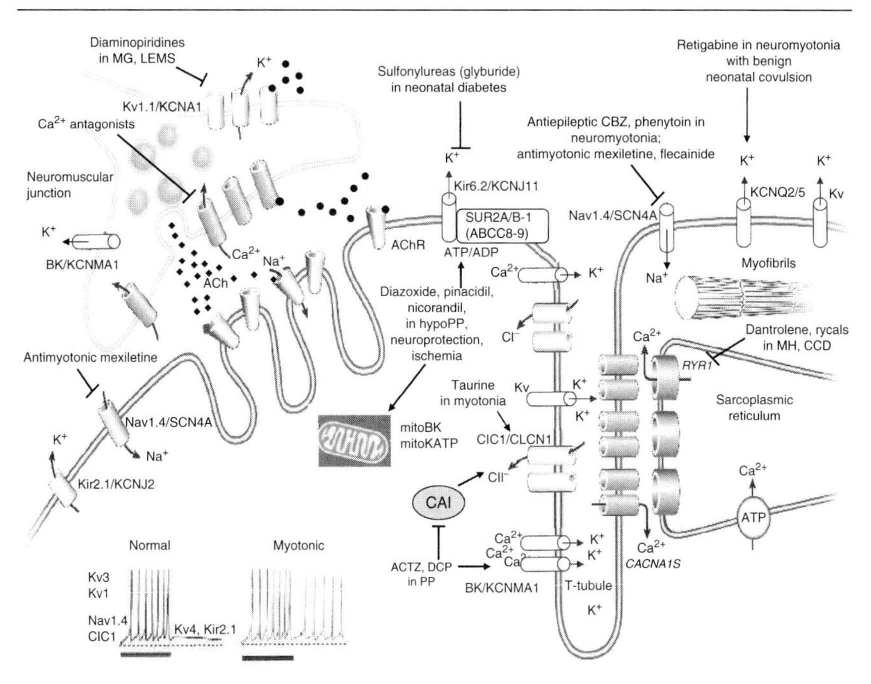

Figure 4.1. Cardiac ion channels involved in channelopathies and targets of drugs. Channelopathies related to cardiac ion channel genes are indicated in plain boxes, while classes of drugs with examples are listed in dashed boxes. AAD, antiarrhythmic drugs. Please refer to the text for other abbreviations.

Steroid hormones, psychostimulants, prokinetics, and other minor drugs are used to correct the large varieties of disease symptoms affecting myotonic dystrophy patients.

1. Periodic paralysis

ACTZ (250–500 mg/day) and DCP (50 mg/day) are first-choice drugs in the treatment of periodic paralyses (PP), which include the hypokalemic and hyperkalemic types, and Andersen–Tawil syndrome (ATS), but are less effective in myotonia (Venance *et al.*, 2006). These drugs reduce the frequency and severity of the paralytic attacks in PP and may ameliorate the muscle strength with various degree of efficacy among disease conditions and patients. The patients affected by PP suffer for episodic attacks of weakness with varying intervals of normal muscle function/excitation and degree of severity. Additional triggers are often required for provoking the muscle paralysis. The hyperkalemic periodic paralysis (hyperPP) and hypokalemic periodic paralysis (hypoPP) are distinguished on the basis of the serum K^+ levels during the attacks of tetraplegia. The intake of K^+ ions and glucose have opposite effects in these disorders. In hyperPP, K^+ ions trigger a hyperkalemic attack, and glucose resolves the attack. Conversely, glucose provokes hypokalemic attacks in hypoPP, which are ameliorated by K^+ ions' intake. In ATS, the main target organ is the heart and the skeletal muscle symptoms are less severe. In this disorder, the paralytic attack may be hyperkalemic or hypokalemic and, accordingly, the response to oral K^+ is unpredictable. Patients with ATS may experience a life-threatening ventricular arrhythmia, and a long QT (LQT) syndrome is the primary cardiac manifestation associated with dysmorphic features.

Drugs targeting KChs are effective in all forms of PP. In hypoPP, KCh openers repolarize the fibers and restore the impaired muscle strength characteristic of this disorder. HypoPP is associated with loss-of-function defects of Cav1.1 (hypoPP type 1) and Nav1.4 (hypoPP type 2), which however do not explain the long-lasting depolarization and weakness seen in patients and the lowering of serum K^+ ions during the attack. A reduced expression/activity of Kir6.2/SUR2A subunits composing the ATP-sensitive K^+ channels (KATP) has been observed in the skeletal muscle of K^+-depleted rats, a secondary form of the disease, and in humans affected by hypoPP type 1 (Tricarico *et al.*, 1998, 1999, 2008b). The KATP channels found in hypoPP patients and in K^+-depleted rats are insensitive to insulin/nucleotide stimulation, thus explaining the insulin/glucose-induced depolarization, paralysis and hypokalemia. KATP channel openers (KCO–KATP) such as cromakalim, diazoxide, and pinacidil have shown efficacy in K^+-depleted rats and in hypoPP patients. Cromakalim repolarizes muscle fibers in hypoPP patients as well as in K^+-depleted rats (Quastoff *et al.*, 1990; Tricarico *et al.*, 1998). Furthermore, pinacidil is effective in reducing the attack frequency and in restoring muscle strength in hypoPP patients (Links *et al.*, 1995). Diazoxide was also found to ameliorate the weakness and paralysis in human hypoPP, possibly by opening skeletal muscle KATP channels and the pancreatic KATP channel subtype, thus reducing the release of insulin. However, first-generation KCO–KATP were not developed against the skeletal muscle KATP channel subtype. They are not skeletal muscle selective and show several side effects such as hypotension, tachycardia, and hyperglycaemia, which limit their use in neuromuscular diseases. New molecules have been found as openers of the skeletal muscle KATP channels (see Section III) (Tricarico *et al.*, 2003, 2008b).

ACTZ and DCP are known for their efficacy in CNS and peripheral channelopathies; however, their mechanism of action is not known. We demonstrated that these drugs exert their effects by activation of the Ca^{2+}-activated K^+ (BK) channel, which is widely expressed in the concerned tissues, including skeletal muscle (Tricarico *et al.*, 2004). These and related drugs prevent fiber depolarization and the paralytic attack induced by insulin in humans and in the K^+-depleted rats, the animal model of hypoPP (Tricarico *et al.*, 2006a). Clinical observations have shown that ACTZ is capable of ameliorating the muscle strength by an unknown mechanism. It is possible that the effect of ACTZ on muscle strength is related with the reported capability of this drug to prevent the characteristic vacuolar myopathy in the K^+-depleted rats by interfering with diffusion and accumulation of lactate from the fibers into the t-tubule (Tricarico *et al.*, 2008a). An abnormally elevated H^+-dependent efflux of lactate from the fibers into the t-tubule is indeed the mechanism responsible for the vacuole formation in K^+-depleted rats and in humans affected by hypoPP and maybe associated with the weakness (Jurkat-Rott and Lehmann-Horn, 2007; Tricarico *et al.*, 2008b). Clinical investigations have shown that ACTZ is effective in hypoPP type 1, but efficacy in hypoPP type 2 is unclear. Alternatively, these patients are treated with triamterene and spironolactone, which resolve the hypokalemia and the paralytic attacks of weakness by restoring the serum K^+ levels. K^+ infusion practice, although in use, is not always effective in ameliorating the attack of weakness and may lead to uncontrolled arrhythmias with atrioventricular block. Drugs targeting Cav1.1 and Nav1.4, such as Ca^{2+} channel (CaCh) antagonists and NaCh blockers, have been tested in hypoPP patients. They were ineffective and in some patients may precipitate the paralysis, thus leading to discontinuation of clinical investigations.

ACTZ and K^+-sparing diuretics are the sole drugs for patients affected by ATS (Sansone and Tawil, 2007). This disease is linked to mutations of the Kir2.1, an inward rectifier K^+ channel expressed in skeletal and cardiac muscle. Kir2.1 channel is essential for maintaining the highly negative resting membrane potential of muscle fibers and accelerating the repolarization phase of the cardiac action potential (AP). The mutations mediate loss-of-channel function by haploinsufficiency or by dominant-negative effects on the WT allele and may lead to long-lasting depolarization and membrane inexcitability. The mutant channel is also unresponsive to the intracellular second messengers. Specific openers of Kir2.1 channels are not available. It is unlikely that the ACTZ efficacy in Andersen's syndrome passes through activation of BK channels, which are not expressed in the heart. It is possible that this drug mostly acts indirectly ameliorating the serum K^+ levels, thus reducing the risk of arrhythmias and LQT syndrome. Alternatively, the intracellular acidification due to CA inhibition by ACTZ may indirectly activate the cardiac KATP with beneficial effect on the AP. Other drugs possibly effective in this disorder are the cardiac KCO–KATP openers, such as nicorandil and pinacidil, which may correct the lack of inward rectifier K^+ currents in cardiomyocytes and AP prolongation; however, no satisfying clinical investigations have been performed with these drugs.

ACTZ is also effective in hyperPP, which results from mutations in the voltage-gated Na^+ channel hNav1.4. Such mutations impair channel inactivation thus leading to a persistent Na^+ current that produces a mild membrane depolarization. Accumulation of K^+ ions into the t-tubule following the electrical activity of the fibers then inactivates most of the Na^+ channels, leading to paralysis. Accordingly, the Nav1.4 blockers as mexiletine, while correcting the myotonia which may be occasionally present in the same patients, are not much effective in preventing weakness in patients with hyperPP and in reducing the frequency of the attacks. The mechanism of action of ACTZ is not directed against NaCh, although lowering of intracellular pH associated to CA inhibition would lead to block of NaChs; it may be possible that modulation of BK in skeletal muscle or in renal and gastrointestinal district play a role, by restoring resting potential and serum K^+ levels. In addition to ACTZ, β_2-adrenoceptor agonists such as salbutamol and terbutaline have shown some efficacy, possibly through activation of the $2Na^+/3K^+$ pump with uptake of K^+ ions into the fibers and repolarization/relaxation of the muscle. A treatment with K^+-loss diuretics such as thiazide derivatives as hydrochlorothiazides is needed to restore the serum K^+ levels in hyperPP attacks.

Few drug treatments have been rigorously studied in neuromuscular channelopathies in a randomized, double-blind CT with the exception of dichlorphenamide in periodic paralysis (Tawil *et al.*, 2000). Currently, the HYP–HOP study is a phase III RCT trial sponsored by the NIH, enrolling patients to determine the efficacy of ACTZ or DCP for treating periodic paralysis and for improving strength [Venance *et al.*, 2007; Hyper- and Hypokalemic Periodic Paralysis Study (HYP–HOP) Clinicaltrial.gov].

2. Nondystrophic myotonias

The drug treatment of the nondystrophic myotonias includes blockers of Nav1.4 channels and openers of Cl^- channels, especially the CLC-1 subtype expressed in skeletal muscle. This is a heterogeneous set of rare diseases that show clinical myotonia, electrical myotonia, or both (Jurkat-Rott and Lehmann-Horn, 2005). These include paramyotonia congenita (PC), potassium-aggravated myotonia (PAM), and the recessive and dominant myotonia congenita (MC). The nondystrophic myotonias are divided into those with Cl^- channel dysfunction such as the myotonia congenital disorders and those with NaCh dysfunction such as paramyotonia congenita, potassium-aggravated myotonia, and periodic paralysis with hyperkalemia and myotonia. This is due to either (1) reduced Cl^- conductance caused by a loss of function of the CLC-1 channel or (2) altered gating kinetics of Nav1.4 channel. Abnormal membrane hyperexcitability occurs as a result of a mild membrane depolarization of few mV, sufficient to recurrently activate Na^+ channels. Although most individuals with mutations in the skeletal muscle Nav1.4 exhibit either a predominant myotonic or periodic paralysis, some have both. The drugs correcting the myotonic symptoms are mexiletine, flecainide, and related drugs, which block voltage-gated NaChs by a use-dependent mechanism. The use-dependent block is indeed the basis for the use of local anesthetic and antiarrhythmic drugs in the treatment of myotonic syndromes, which are characterized by high-frequency AP discharges in skeletal muscles. Other than mexiletine, which is marketed as an antiarrhythmic drug, other NaCh blockers used against myotonia include tocainide, procainamide, disopyramide, phenytoin, flecainide, and carbamazepine. Mexiletine is administered in doses of 150–200 mg two to three times a day and is generally well tolerated. Cardiac function and drug serum concentration should be carefully monitored, to reduce the risk of CNS and cardiac toxicity. These drugs are effective in patients carrying either the CLC-1 or Nav1.4 mutations; however, they are not effective in the treatment of periodic paralysis associated with hyperkalemia. The mutations in Nav1.4 may modify channel sensitivity to mexiletine, which might be exploited for pharmacogenetic studies (Desaphy *et al.*, 2001, 2004). New synthetic molecules targeting specific mutants have been investigated by us and others (see Section III).

Drugs of interest for myotonia congenita should be able to increase Cl^- currents, since mutations produce a loss of function of the encoded CLC-1 channel. Identification of specific CLC channel openers would be useful for counteracting hyperexcitability in myotonia. The other way to enhance Cl^- currents is to indirectly increase CLC channel expression or function by stimulating intracellular biochemical pathways. For instance, it is noteworthy that chronic treatment with taurine may improve myotonia, while the amino acid is known to increase the muscle Cl^- conductance *in vivo* and *in vitro* (Conte Camerino *et al.*, 2004). More recently, ACTZ was shown to increase CLC-1 channel currents in mammalian cell lines, most probably through intracellular acidification (Eguchi *et al.*, 2006). Thus, the indirect activation of CLC-1 channels might also be useful for treating myotonia.

3. Myotonic dystrophies

The pharmacotherapy of myotonic dystrophies involves the use of different classes of drugs. This is indeed a group of autosomal-dominant multiorgan diseases, which include myotonic dystrophy type 1 (DM-1), proximal myotonic myopathy/myotonic dystrophy type 2 (PROMM/DM-2), and proximal myotonic dystrophy (a variant of DM-2) (Heatwole and Moxley, 2007). Myotonia is one among the many neurological, gastrointestinal, ocular, and cardiac symptoms of these progressive diseases. One of the most severe symptoms is muscle wasting and atrophy. The mutations in myotonic dystrophies consist in a toxic expansion of unstable polynucleotide repeats in specific DNA regions. The pathogenesis in DM-1 and DM-2 is thus different from that of the nondystrophic myotonias. However, toxic microsatellite expansion of mRNA may lead to altered expression/splicing of various proteins, including the CLC-1 channel and the SERCA

Ca^{2+}-ATPase in skeletal muscles, and the cardiac NKX2-5 connexins (Yadava *et al.*, 2008). Thus, a reduction in the major skeletal muscle Cl^- channel (CLCN1) may be responsible for the myotonic syndrome. The drug therapy is focused to resolve the various severe symptoms of these disorders by using drugs such as prokinetic as bethanechol, antidyspeptic drugs and laxatives, which are aimed at correcting the gastrointestinal motility disorder, and dehydroepiandrosterone to reduce inflammation. Psychostimulants such as modafinil are used to counteract the excessive daytime sleepiness, which is commonly observed in the myotonic dystrophy (Wintzen *et al.*, 2007). Mexiletine and taurine are known for their ability to correct myotonia. We also proposed clenbuterol, a β_2-adrenoceptor agonist, for the possibility to combine its well-known anabolic action with its antimyotonic activity through NaCh use-dependent block, but clinical data are lacking (Desaphy *et al.*, 2003). Some clinical investigations were performed to establish the efficacy and safety of drugs in patients affected by myotonic dystrophies and myotonia congenita. Nine randomized controlled trials were found comparing active drug treatment versus placebo or another active drug treatment in patients with myotonia due to a myotonic disorder (Trip *et al.*, 2006). Those trials were double-blind or single-blind crossover studies involving a total of 137 patients of which 109 had myotonic dystrophy type 1 and 28 had myotonia congenita. Two small crossover studies demonstrated a significant effect of imipramine and taurine in myotonic dystrophy. One small crossover study demonstrated a significant effect of clomipramine in myotonic dystrophy. Single studies give an indication that chlorimipramine and imipramine have a short-term beneficial effect and that taurine has a long-term beneficial effect on myotonia (Annane *et al.*, 2006; Trip *et al.*, 2006). Larger, well-designed randomized controlled trials are needed to assess the efficacy and tolerability of drug treatment for myotonia.

4. Malignant hyperthermia and central core disease

MH is an autosomal-dominant predisposition to respond abnormally when exposed to volatile anesthetics, depolarizing muscle relaxants, or extreme physical activity in hot environments, with acute and fatal crisis being a challenge for anesthesiologists (Bellinger *et al.*, 2008; Jurkat-Rott and Lehmann-Horn, 2005). During exposure to triggering agents, a pathologically high increase in myoplasmic Ca^{2+} concentration leads to increased muscle metabolism and heat production, resulting in muscle contractures, hyperthermia associated with metabolic acidosis, hyperkalemia, and hypoxia. The metabolic alterations usually progress rapidly, and, without immediate treatment, up to 70% of the patients die. Early administration of dantrolene, an inhibitor of Ca^{2+} release from the SR, has successfully aborted numerous fulminant crises and has reduced the mortality rate from 80% in the 1960s to <10% today (Krause *et al.*, 2004). Dantrolene is a hydantoin derivative known from many years for its depressive action on the EC coupling. This drug is available for intravenous use. Dantrolene analogue such as azumolene has been prepared and investigated for its effect in MH; however, no advantage emerged with respect to dantrolene, which remains the sole drug available for MH. Dantrolene and azumolene discriminate between the skeletal muscle and cardiac ryanodine receptor (RyR) subtypes, but the exact location of the binding site has not been identified. In most families, mutations can be found in genes encoding the skeletal muscle ryanodine receptor (RyR1) or voltage-gated Ca^{2+} channel (Cav1.1). RyR1 is a Ca^{2+}-releasing channels that is not voltage dependent on its own, but under the control of Cav1.1 channel. The MH mutations are usually located in the cytosolic part of the protein and show gain-of-function effects: they increase RyR1 sensitivity to caffeine and other activators, as shown in functional tests of excised muscle, isolated native proteins, and ryanodine receptors expressed in muscle and nonmuscle cells.

Mutations in the SR-luminal region of RyR1 cause CCD, a congenital myopathy clinically characterized by muscle hypotrophy and weakness and a floppy-infant syndrome, often accompanied by other skeletal abnormalities such as hip displacement and scoliosis. Pathognomonic is the abundance of central cores lacking oxidative enzyme activity along the predominant type 1 muscle fibers. Mutations increase the open probability of the RyR1 channel, leading to depleted SR Ca^{2+} stores and weakness. Both dominant and rare recessive mutations have been described, the latter transiently presenting as multiminicore disease. The SR Ca^{2+} release channels form an enormous macromolecular signaling complex comprising four approximately 565-kDa RyR1 monomers and several enzymes that are targeted to the cytoplasmic domain of the RyR1 channel, including calstabin1. Calstabin1 stabilizes the channel in the closed state conformation and modulation of this enzyme is a key step in controlling channel activity, thus constituting a valuable drug target. Such a drug is known as *rycals*, which inhibit the SR Ca^{2+} leak by binding to calstabin1 on RyR1 complex. This compound reduces myofiber damage, sarcopenia, and death associated with stress-signaling pathways targeting RyR1. Furthermore, preventing the dissociation of calstabin1 from the RyR1 complex and fixing the RyR1 channel leak in skeletal muscle using novel small molecules can improve muscle fatigue in mice with heart failure. However, drugs for clinical use are not available.

B. Cardiac channelopathies

Mutations of ion channels expressed in the heart result in life-threatening cardiac arrhythmias. Management of arrhythmia with antiarrhythmic drugs is not an easy issue. Among antiarrhythmic drugs, only the β-blockers (class II antiarrhythmics according to the Vaughan Williams classification) have been shown to reduce sudden cardiac death in randomized clinical trials (Zipes *et al.*, 2006). When β-blockers therapy is not effective, the KCh blockers (class III), amiodarone or sotalol, can be tried with monitoring for side effects. The other antiarrhythmic drugs include NaCh blockers (class IA, IB, and IC) and CaCh antagonists (class IV). These drugs can be used under certain circumstances and in particular types of arrhythmia. In the case of cardiac channelopathies, there is a continuous effort to search for efficient mechanism-based therapeutic strategies relative to each disease (Fig. 4.2).

Inheritance of the monogenic cardiac channelopathies is either autosomal dominant or recessive, depending on the specific mutation (Lehnart *et al.*, 2007). The genetic LQT syndromes are due to gain of function of NaChs and CaChs, or loss of function of KChs. The channel dysfunction augments ventricular AP duration, which appears as an increased QT interval on the electrocardiogram (ECG). This increases the propensity for early afterdischarges, which in turn determine atypical polymorphic ventricular tachycardia known as *torsade de pointes* and a high risk of sudden cardiac death. A consistent proportion of cases of sudden infant death syndrome (SIDS) are caused by LQT syndromes (Arnestad *et al.*, 2007). Conversely, gain-of-function gene mutations in KChs or loss-of-function mutations of CaChs lead to enhanced repolarization with shortening of the QT interval, which may provide a substrate for re-entrant arrhythmias in the short QT (SQT) syndromes. Loss-of-function mutations of *SCN5A* encoding Nav1.5 channels have been linked to Brugada syndrome, familial atrial fibrillation, sick sinus syndrome, and cardiac conduction diseases. In addition, a mutation in the NaCh auxiliary β_1-subunit (*SCN1B*) was recently found to lead to Brugada syndrome and cardiac conduction defect (Watanabe *et al.*, 2008). Mutations in the *HCN4* gene encoding the pacemaker channel have been associated with sick sinus syndrome, atrial fibrillation, and bradycardia. Finally, mutations in *RyR2* gene encoding the intracellular ryanodine-sensitive Ca^{2+} release channel have been linked to catecholaminergic polymorphic ventricular tachycardia (CPVT).

Physical exercise and emotional stress often precipitate syncope in patients with idiopathic LQT (Shimizu *et al.*, 2005). Thus, the β-blockers, by inhibiting catecholaminergic stimulation, represent the preferred current drug therapy for most patients with either LQT or CPVT. However, several studies have shown that LQT2 and especially LQT3 are quite resistant to β-blockers, in accordance with the lower incidence of syncope during exercise as compared with LQT1. Clinical and experimental studies advocate the use of the class IB NaCh blocker, mexiletine, which is able to inhibit the aberrant long-lasting Na^+ current and to shorten QT interval in LQT3 patients (Schwartz *et al.*, 1995; Wang *et al.*, 1997). The class IC NaCh flecainide has also been considered for its specific effect on a subset of LQT-causing NaCh mutations, but the drug may increase the likelihood for LQT3 patients to present symptoms reminiscent to

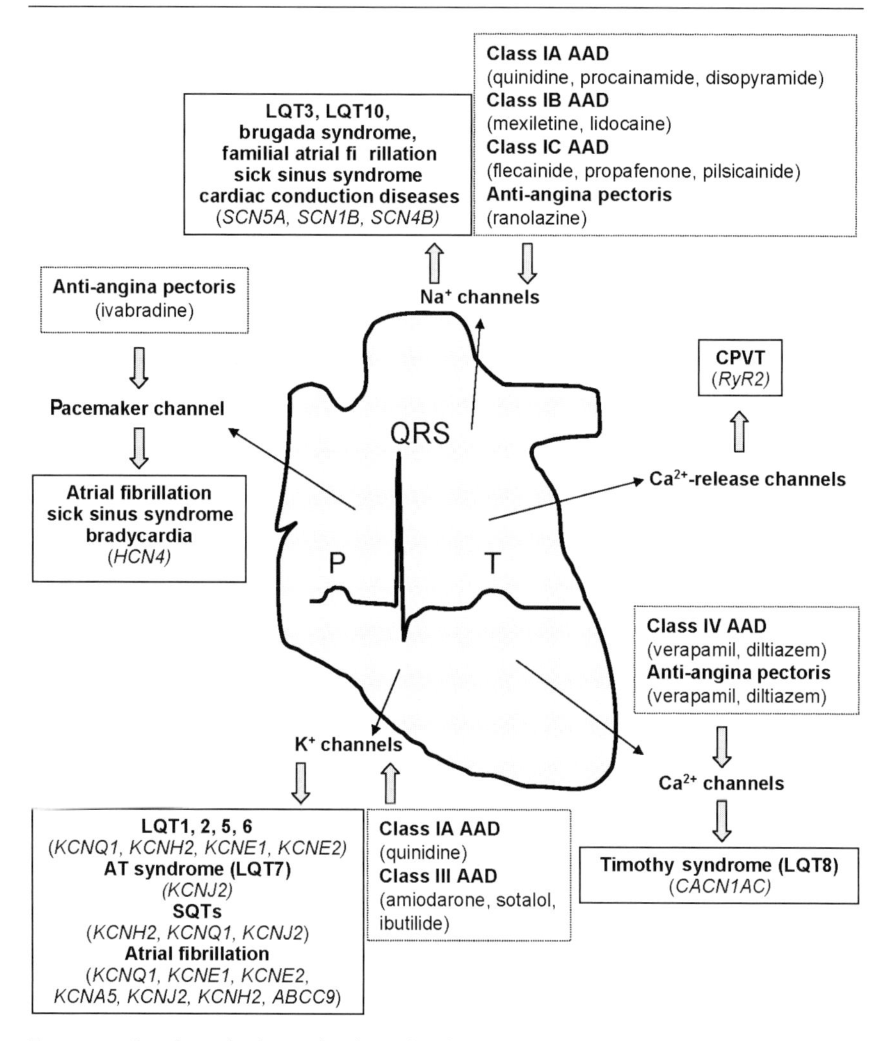

Figure 4.2. Ion channels of central and peripheral nervous system involved in channelopathies and targets of drugs. Channelopathies related to ion channel genes are indicated in plain boxes, while classes of drugs with examples are listed in dashed boxes. VG, voltage gated. Please refer to the text for other abbreviations.

Brugada syndrome (Benhorin *et al.*, 2000; Priori *et al.*, 2000). The use of Ca^{2+} antagonists (e.g., verapamil) and KCh openers (e.g., nicorandil) in LQT treatment was also proposed, but there are little clinical data available (Shimizu *et al.*, 2005). In addition to pharmacotherapy, high-risk patients with a strong personal history of syncope and marked QT prolongation may receive an ICD device. Left cardiac sympathetic neural denervation is also proposed in drug-resistant individuals. The ICD device is the first-choice therapy of high-risk patients with Brugada syndrome or SQT syndrome. A number of antiarrhythmic drugs, including β-blockers, class IC NaCh blockers, and CaCh antagonists, produce an elevation of ST segment on the ECG, which is characteristic of Brugada syndrome and associated to syncope or sudden death outcome. Thus, these drugs are not expected to work in Brugada syndrome, while class IC drugs are used as a diagnostic tool to unmask latent Brugada syndrome. Alternative pharmacotherapeutic tentative has been performed with the class IA drug quinidine for its ability to block various KChs (Belhassen *et al.*, 2004). Quinidine may also be useful in some SQT patients, although others are not responsive due to drug-resistant polymorphisms in KCh genes. Thus, there is an elevated variability of arrhythmic patient response to either treatment in terms of sudden cardiac arrest prevention and side-effect tolerance. Moreover, the therapy may be quite complicated in case of mutations associated with mixed clinical phenotype (Makita *et al.*, 2008). It is therefore evident that development of effective novel mechanism-based therapeutic strategies is thoroughly needed. In cases of loss-of-function gene mutations causing protein trafficking abnormalities, drugs able to correct or improve trafficking may offer a new therapeutic approach. On the other hand, gating modifying drugs would be useful in cases of gene mutations altering channel gating. This implicates the availability of facile, routine mutation screening test, which will allow us to select the best drug for each patient/mutation. In addition, gene and cell therapies are under development, which require preclinical evaluation of efficacy and safety.

C. Neurological channelopathies

1. Idiopathic epilepsy

Epilepsy regroups a large number of different diseases characterized by recurrent seizures, which can cause motor, sensory, cognitive, psychic, or autonomic disturbances (Steinlein, 2004). The seizure is determined by a transient hyperexcitability of cortical neuron networks, while the absence seizure results from an abnormal synchronization of the thalamocortical circuit. By convention, epilepsy diagnosis is applied for individuals who have experienced at least two seizures. Recurrent seizures can alter neuronal networks, inducing permanent damage and increasing the risk of seizure occurrence.

Idiopathic epilepsies can display all possible modes of inheritance, including autosomal, X-linked, mitochondrial, and complex. Polygenic epilepsies are much more common that monogenic forms, but the latter have significantly contributed to our understanding of the pathogenesis. Many idiopathic epilepsies have been associated to mutations in genes encoding ion channels or accessory subunits (Cossette *et al.*, this volume). Generally, neuronal hyperexcitability can easily be conciliated with the gain or loss of ion channel function induced by mutations. Nevertheless, the phenotype–genotype relationships may remain a matter for debate, as for NaCh-associated epilepsies (Ragsdale, 2008). It is noticeable that many of the available antiepileptic drugs (AEDs) act by modulating ion channel function (Fig. 4.3).

About half of the patients can manage seizures by using a single AED. A single drug is preferred to combine therapy in order to limit costs, occurrence of side effects, and interactions between drugs. The choice of the AED is based first, on seizure type and second, on physical condition and concomitant medications. Initially, the patient receives a low dosage, which is gradually increased up to the maximum-tolerated dose. If not successful or in case of intolerance, the patient will receive the other AED. If still not sufficient, the patient may receive a combination of AEDs. Over the last 15 years, a large arsenal of AEDs has been made available and numerous other compounds are under preclinical or clinical evaluation (Perucca *et al.*, 2007). This arsenal includes positive modulators of GABA (γ-aminobutyric acid) neurotransmission, inhibitors of voltage-gated NaChs and CaChs, drugs with multiple mechanisms, and empiric drugs whose exact mechanism of action remains uncertain.

GABA is the main inhibitory neurotransmitter in the brain. Binding to $GABA_A$ receptor type on the postsynaptic membrane opens a Cl^- channel, which leads to membrane hyperpolarization and transmission inhibition. Phenobarbital was the first AED available on the market since 1912. It works against partial-onset seizures, primary generalized tonic-clonic seizures, and myoclonic seizures. Endovenous phenobarbital can also be useful against status epilepticus, a life-threatening state characterized by recurrent or continuous seizure activity lasting longer than 30 min. Like other barbiturates, it binds to $GABA_A$ receptor close to the channel pore and enhances GABA action. Although doses needed for anticonvulsant action are lower than ipnotic doses, phenobarbital can induce sedation and behavioral effects. Moreover, it is a potent inductor of metabolism. It is thus considered as a second choice today. Currently available is also primidone, which is a prodrug of phenobarbital and displays merely the same effects. The benzodiazepines (BZDs: diazepam, clonazepam, clobazam, lorazepam, nitrazepam, midazolam) act similarly to phenobarbital but bind to a different site located on the extracellular side of $GABA_A$ receptors. Benefits and pitfalls of BZDs are similar to those of phenobarbital. Among the more recent AEDs, tiagabine is a derivative of nipecotic acid, which inhibits neuronal and glial reuptake of the neurotransmitter, thereby increasing GABA availability in the synaptic cleft. Tiagabine is effective against partial seizures but may aggravate absence seizures. It is metabolized by hepatic enzymes and is thus sensitive to metabolism modifiers. Other drugs are known to modulate GABA transmission as part of their anticonvulsant action, including valproate, topiramate, and felbamate.

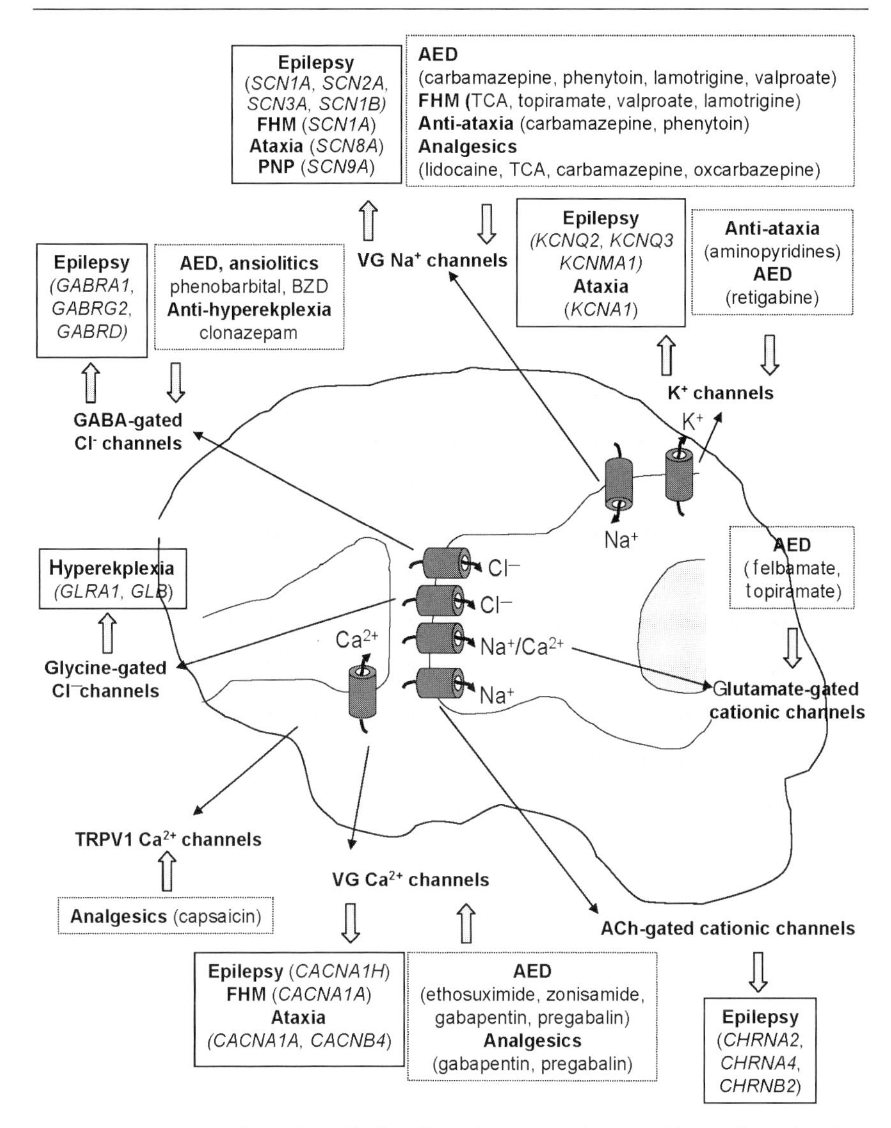

Figure 4.3. Kidney ion channels involved in channelopathies and targets of drugs. Channelopathies related to ion channel genes are indicated in plain boxes, while classes of drugs with examples are listed in dashed boxes. Please refer to the text for abbreviations.

The other mechanism to decrease neuron hyperexcitability consists in blocking voltage-gated NaChs, which are responsible for AP upstroke. Sodium channel blockers (phenytoin, carbamazepine, oxcarbazepine, lamotrigine) are first-line treatment for partial-onset and generalized tonic-clonic seizures. They do not work against myoclonic seizures and absence seizures. Other AEDs, including valproate, felbamate, topiramate, and zonisamide, are able to block NaChs, which may contribute to their anticonvulsant activity. It is thought that their molecular receptor matches, at least partially, that of the local anesthetics sited within the ion-conducting pore (Ragsdale *et al.*, 1996). Recently, a common *SCN1A* polymorphism was shown to influence the clinical use of phenytoin and carbamazepine (Tate *et al.*, 2005). Moreover, a new *SCN3A* mutation was identified in a patient with carbamazepine-resistant cryptogenic pediatric partial epilepsy (Holland *et al.*, 2008). Also, an epileptic mutation in the auxiliary β_1-subunit (mutation C121W in *SCN1B*) was shown to reduce *in vitro* the sensitivity of the mutant channel to phenytoin (Lucas *et al.*, 2005). In addition, many *SCN1A* mutations may produce a loss of function of the NaCh (Ragsdale, 2008), suggesting that NaCh blockers may be inappropriate in patients carrying these mutations. Importantly, such results open the way for the development of effective pharmacogenetics.

The voltage-gated Cav3 channels are expressed in somatic and dendritic areas of neurons in the hippocampus, hypothalamus, thalamus, cerebellum, and cortex, being responsible for the low-threshold T-type Ca^{2+} currents. Abnormal openings of these channels are the basis of the low-frequency discharge (3 Hz) of hypothalamic nucleus associated with absence epilepsy. A genetic association study also proposed *CACNA1H*, which encodes the Cav3.2 α_1-main subunit, as a susceptibility gene involved in childhood absence epilepsy (Chen *et al.*, 2003). These channels are the main target of ethosuximide, which is a first-line treatment of absence seizures. Other AEDs having a multiple mode of action, such as zonisamide and valproate, can also inhibit T-type CaChs and are effective against absence seizures.

More recently, two GABA analogues, gabapentin and pregabalin, were unexpectedly found to act primarily by targeting the $\alpha_2\delta$-CaCh auxiliary subunit (Dooley *et al.*, 2007). The $\alpha_2\delta$-subunit associates with the ion-conducting pore α_1-subunit in a 1:1 stoichiometry. Several genes are known to encode different $\alpha_2\delta$- and α_1-subunits, and numerous splice variants have been found. There is no clear description of which $\alpha_2\delta$- and α_1-subtypes coassemble together. The $\alpha_2\delta$-ligands are effective against partial-onset seizures but may exacerbate myoclonic and absence seizures. This suggests that, unlike ethosuximide, they do not act on T channels, but rather on N and/or P/Q CaChs (Cav2 type), which are presynaptic channels involved in the release of various neurotransmitters. Such targets may explain the potential wide application of gabapentin, which received approval also for treatment of neuropathic pain and generalized anxiety disorders.

Two other recent AEDs may exert, at least in part, their anticonvulsant activity by inhibiting excitatory glutamate transmission. Felbamate antagonizes *N*-methyl-D-aspartate (NMDA-type) glutamate receptors, while topiramate may antagonize non-NMDA (probably AMPA-type) receptors. They are both used for partial-onset and primary generalized tonic-clonic seizures. In addition, felbamate may be effective against myoclonic seizures.

Nonion channel targets of AEDs include synaptic vesicle proteins and carbonic anhydrase enzymes. Among AEDs, levetiracetam shows a unique pharmacodynamic profile. By binding to the presynaptic vesicle protein SV2A, it may impair release of various neurotransmitters. Effects on GABA and glycine-activated Cl^- currents and on K^+ currents were also proposed to contribute to anticonvulsant action of levetiracetam. The drug is effective against partial-onset and generalized tonic-clonic seizures, as well as myoclonic and absence seizures. Interestingly, levetiracetam add-on therapy was recently shown to be effective against severe myoclonic epilepsy of infancy (SMEI or Dravet syndrome), an idiopathic epilepsy characterized by frequent, drug-resistant seizures of many types occurring between ages 1 and 4, which impair psychomotor development (Striano *et al.*, 2007). Some SMEI patients were found to carry loss-of-function mutations in the *SCN1A* NaCh, which might explain SMEI resistance to Na^+ channel blockers.

Some carbonic anhydrase inhibitors, including ACTZ, methazolamide, and sulthiame, have been empirically used for a long time against various types of seizures, either in combination therapy or in refractory epilepsy (Thiry *et al.*, 2008). The exact mechanism of action of such compounds in epilepsy is quite unclear. Modulation of pH in the CNS, which in turn modulates ion channel function at the synapse, may reduce neuronal excitability. Moreover, some CAI including ACTZ is recently shown to activate large conductance, Ca^{2+}-activated K^+ channels, which might contribute to neuronal inhibition (Tricarico *et al.*, 2004). The second-generation AED, zonisamide, inhibits carbonic anhydrase *in vitro* and was expected to suppress seizures through CA inhibition. It appeared, however, that its anticonvulsant activity is mainly attributable to inhibition of NaChs and T-type CaChs. On the other hand, CA inhibition was unexpectedly shown to contribute to anticonvulsant activity of topiramate. Thus, CA inhibition may represent a powerful target for development of new AEDs (Thiry *et al.*, 2008).

2. Familial hemiplegic migraine

Migraine is a chronic neurologic disorder characterized by repeated headache attacks, nausea, and an increased reactivity to sensory stimuli. Aura may be present or not. A low migraine threshold is set by genetic factors and modulated by phenotype. Migraine is a relatively frequent disorder affecting 6–9% individuals among men and 15–17% among women. Conversely, familial hemiplegic migraine (FHM) is a rare monogenic disease. For instance, prevalence was found to be ~1/20,000 individuals in Denmark (Lykke Thomsen *et al.*, 2002).

Mutations in *CACNA1A*, *ATP1A2*, and *SCN1A* have been associated with FHM, which all encode ion channels/carriers. The mutations described so far encompass only a limited number of FHM kindred and other genomic regions have been linked to the disease. This recent knowledge has been useful in understanding the migraine pathogenesis, but has had little repercussion on therapy yet, which is identical for polygenic migraine and FHM.

It is well established that direct treatment costs are much lower than the economical cost of absenteeism or reduced work productivity of individuals during attacks. First-choice drugs for acute treatment of migraine attacks include NSAIDs (nonsteroidal anti-inflammatory drugs) (e.g., acetylsalicylic acid, ibuprofen, diclofenac, paracetamol) to relief pain and dopamine antagonists (metoclopramide, prochlorperazine, chlorpromazine) to treat nausea, which are the most troublesome symptoms during the attacks. Although not recommended, opioids are also largely used to treat migraine in the emergency department (Colman *et al.*, 2004). Ergot alkaloids (e.g., dihydroergotamine, ergotamine/caffeine) have been used for a long time. Their efficacy in migraine attacks is probably related to partial agonist activity toward serotonin receptors. However, they are nonselective agents and may induce severe side effects. They are less efficient than triptans (e.g., sumatriptan, zolmitriptan, naratriptan, rizatriptan, etc.), which are selective 5-HT1 receptor agonists (Christie *et al.*, 2003; Diener *et al.*, 2002). Activation of these receptors may help resolving migraine by producing contraction of intracranial vessels and by reducing the release of proinflammatory neuropeptides. Introduction of triptans during the 1990s has noticeably ameliorated the treatment of migraine. These drugs are generally well tolerated, although cardiovascular undesirable events may be observed in predisposed patients.

The troubles caused by migraine may not be limited to the attack episode. Some migraineurs may suffer from poor well-being and anxiety due to the unpredictable occurrence of attacks and prevention behavior. In these individuals as well as in those suffering from several attacks per month or not responding to acute therapy, a prophylactic drug treatment can be initiated. Preventive treatment can reduce the frequency, duration, or severity of attacks and improve responsiveness to acute attack. This reduces disability and the cost of health care. The most widely used drugs are the β-adrenoceptor blockers (e.g., propranolol), tricyclic antidepressants (e.g., amitriptyline), anticonvulsants (e.g., topiramate and valproate), and serotonergic drugs (e.g., methysergide). The older medications (propranolol and amitriptyline) were found by chance to prevent migraine in patients treated for other conditions, whereas the use of topiramate results from prospective studies based on prior success of some anticonvulsants. There are evidences that suppression of cortical spreading depression, characterized by 1-min-long depolarization of neurons and glial cells in some brain areas, may be a common endpoint for many prophylactic drugs, especially in migraine with aura (Ayata *et al.*, 2006). However, the fine mechanisms by which these drugs decrease the frequency of migraine attacks are unclear because they can act on a large number of different targets, including several voltage-gated and neurotransmitter-gated ion channels (Fig. 4.3) (Silberstein, 2006). For example, topiramate modulates the activity of NaChs and CaChs as well as GABA and glutamate-gated channels, through prevention of protein kinase A-dependent phosphorylation of channel subunits. In addition, it inhibits a number of CA isoenzymes. Because transient increase in glutamate has been measuring during cortical spreading depression, it is reasonable to propose inhibition of glutamate neurotransmission as one of the action mechanisms of topiramate, while block of NaChs may contribute to antinociceptive effects. Accordingly, lamotrigine, which inhibits glutamate release by blocking NaChs, was shown effective in preventing migraine with aura in a few studies. Because gain-of-function mutations of P/Q-type CaChs result in FHM, one may expect benefits from the use of P/Q channel blockers. However, no conclusive result has emerged from clinical trials with gabapentin, which binds two of the four $\alpha_2\delta$-subunit isoforms, and the drug is not currently approved for such treatment. The potential of the other $\alpha_2\delta$-ligand, pregabalin, in the prophylaxis of migraine would deserve randomized clinical trials.

3. Ataxia

The episodic ataxias (EA) are rare inherited syndromes characterized by intermittent cerebellar dysfunction in individuals with essentially normal brain function. To date, there are six genetically defined types of EA, all of them autosomal dominant, while some other similar diseases still await genotyping. Mutations have been found in genes encoding the voltage-gated Kv1.1 KCh (EA1), P/Q-type CaCh α_1- and β_4-subunits (EA2, EA5), and a glutamate transporter (EA6). In addition, a protein truncation mutation of the NaCh *SCN8A* gene was recently found in a patient with cerebellar atrophy, ataxia, and mental retardation (Trudeau *et al.*, 2006). Often, patients with EA display other symptoms classically related to ion channel dysfunction, including neuromyotonia, seizures, and migraine headaches. Interestingly, mutations in EA2 induce loss of function of P/Q-type CaChs, whereas the allelic mutations responsible for FHM induce a gain of function of the same channel (Strupp *et al.*, 2007). Additionally, CAG repeat expansion in the *CACNA1A* gene has been associated with a closely related disorder called spinocerebellar ataxia type 6 (SCA6) (Kordasiewicz and Gomez, 2007).

First-choice treatment of EA1 includes the anticonvulsants, carbamazepine and phenytoin, which prevent seizures and improve ataxia and myokymia (Fig. 4.3). Response to ACTZ is variable among EA1 patients (Rajakulendran *et al.*, 2007). The patients with EA2, EA3, EA5 as well as those with various nongenetically defined EA respond generally well to ACTZ. The benefits of ACTZ in EA2 were discovered accidentally in a patient misdiagnosed with periodic paralysis. Indeed, ACTZ prevents or reduces the attacks in more than half of the patients, although many have to stop ACTZ after a certain time, because the drug becomes no more effective or the patient becomes intolerant. The mechanism of action of ACTZ in EA likely involves modulation of ion channels by reducing intracellular pH through CA inhibition. Although ACTZ remains the first-choice drug, the CA inhibitor sulthiame may have fewer side effects. ACTZ has been anecdotally shown to improve ataxia also in SCA6 patients, but there is currently no well-defined therapy. The anticonvulsants, gabapentin and pregabalin, which target the auxiliary $\alpha_2\delta$-subunit of P/Q channels, were recently hypothesized to be effective against SCA6, but no clinical data are available yet (Gazulla and Tintoré, 2007). On the other hand, the KCh blocker, 4-AP, was tested in EA2 patients no more responsive to ACTZ. The drug fully prevented attacks in most of the patients, while drug removal led to symptoms recurrence. The drug may increase the release of GABA by Purkinje cells by blocking different kinds of KChs. The aminopyridine may thus represent a new therapeutic alternative in EA2 and maybe other episodic ataxia. For instance, their therapeutic potential may deserve attention in EA4 patients, who do not respond to ACTZ.

4. Hyperekplexia

Hyperekplexia or hereditary startle disease is a disorder characterized by excessive startle responses provoked by acoustic or tactile stimuli (Zhou *et al.*, 2002). It is also know as stiff-baby or stiff-man syndrome, because generalized stiffness is a particularly evident symptom. Hyperekplexia is a rare disorder caused by loss-of-function mutations in the *GLRA1* and *GLB* genes encoding subunits of the glycine-gated Cl^- channel expressed in spinal cord and brainstem, thereby resulting in hyperexcitability (Bakker *et al.*, 2006). Mutations in other genes (*GPHN*, *ARHGEF9*, *SLC6A5*) have been found in some cases, which encode proteins involved in glycine receptor clustering or synaptic glycine concentration control.

The involuntary jerking characteristic of hyperekplexia is considered a reticular reflex myoclonic seizure. Thus, trials with AEDs, including benzodiazepines (diazepam, nitrazepam), phenobarbital, Na^+ channel blockers (carbamazepine, phenytoin, phenytoin), and valproic acid, have been performed with variable results. Treatment with BDZs, especially clonazepam, alone or in combination with the other AED is the mainstay therapy. Clonazepam at high doses dramatically reduces startle responses, is well tolerated, and remains effective on the long run. By enhancing GABA-gated Cl^- currents, benzodiazepines most probably compensate for the loss of function of glycine-gated Cl^- channels.

Recently, preclinical studies suggested that subanesthetic doses of propofol may be able to counteract hyperekplexia mutations by enhancing glycine receptor response (O'Shea *et al.*, 2004). Clinical trials are warranted to verify benefits of propofol compared with clonazepam in hyperekplexia patients.

5. Peripheral pain disorders

Sources of persistent pain may be nociceptive or neuropathic. While nociceptive pain is considered as a physiologic, protective response of the organism, neuropathic pain is defined as pain initiated or caused by a primary lesion or dysfunction in the nervous system. Idiopathic peripheral pain disorders (PPDs) enter the definition of neuropathic pain. Very recently, the genetic cause of three inherited PPDs was undiscovered in the *SCN9A* gene encoding the Nav1.7 NaCh, which is expressed mainly in dorsal root ganglions (Drenth and Waxman, 2007). Gain-of-function mutations have been associated with primary erythermalgia (PE) and paroxysmal extreme pain disorder (PEPD), whereas loss-of-function mutations have been associated with channelopathy-associated insensitivity to pain. Noteworthy, pain is also a prominent symptom of FHM due to mutations in the P/Q-type CaCh.

Although pain may be the more common complaint of individuals consulting physicians, satisfactory treatment of chronic pain remains a major challenge in clinical medicine. A battery of drugs are available for pain sufferers, although the response may vary significantly between individuals and between neuropathic pain types. Expert guidelines have been recently published, based on the review of literature (Attal *et al.*, 2006; Dworkin *et al.*, 2007). For instance, first-line drugs recommended by the European Federation of Neurological Societies include tricyclic antidepressants (TCA), AEDs (pregabalin, gabapentin, carbamazepine, oxcarbazepine), and topical lidocaine (Attal *et al.*, 2006). Drugs with moderate activity, unfavorable efficacy-to-tolerability ratio, or insufficient randomized controlled trials were considered as second-line drugs, including some AEDs (lamotrigine, topiramate, valproate), tramadol, capsaicin, cannabinoids, serotonin–noradrenaline reuptake inhibitors (SNRI), and opioid agonists. The recommendations of the International Association for the Study of Pain show little differences (Dworkin *et al.*, 2007).

The TCA (nortriptyline, clomipramine, desipramine, amitriptyline, etc.) and SNRI (duloxetine and venlafaxine) may act primarily through inhibition of noradrenaline reuptake, whereas selective inhibitors of serotonin reuptake exhibit insufficient pain relief. Their use is limited due to side effects including cardiac QT prolongation. Use-dependent block of NaChs may also contribute to TCA mechanism of action. Topical lidocaine blocks NaChs in a use-dependent manner and thus inhibits ectopic discharges. It is effective against postherpetic neuralgia with minimal systemic adsorption and very few side effects. Carbamazepine and oxcarbazepine are two NaCh blockers commonly used in trigeminal neuralgia. Other NaCh-blocking AEDs may be efficient analgesics, but they are considered as second-choice drugs. The $\alpha_2\delta$-CaCh ligands, gabapentin and pregabalin, were developed for treatment of epilepsy and proved benefits in the treatment of neuropathic pain. Presumably, analgesic action involves blocking of presynaptic CaChs, which reduces glutamate release from primary afferent fibers in the spinal cord. Efficacy of opioids in neuropathic pain is debated, and their use is limited by side effects, development of tolerance, possible addiction, and government policies. Tramadol is a weak agonist of opioid μ-receptor and inhibits reuptake of monoamines. The rationale for topical use of capsaicin is that such compound inhibits nociceptive neurons, by activation and further desensitization of Ca^{2+}-permeable TRPV1 channels. Its use is, however, limited by side effects at the site of application (burning sensation and irritation).

Primary familial erythermalgia is notably highly resistant to treatment. In the specific case of PPD due to gain-of-function Nav1.7 mutations, one may expect beneficial effects of NaCh blockers. Accordingly, lidocaine and mexiletine treatment was shown to be effective in PE patients (Legroux-Crespel *et al.*, 2003). Nevertheless, carbamazepine is notably less effective in PE patients, whereas the drug is able to control the symptoms in some PEPD patients (Drenth and Waxman, 2007). Clearly, more studies are needed to address this issue. Such studies should ideally include genotyping and clinical testing of drugs, as well as *in vitro* evaluation of drug effects on NaCh mutants. For instance, carbamazepine was shown to block heterologously expressed NaChs carrying the same mutation as the PEPD patients responsive to the drug (Fertleman *et al.*, 2006). On the other hand, a *SCN9A* mutation responsible for PE was shown to reduce lidocaine sensitivity (Sheets *et al.*, 2007). Once again, such results highlight the therapeutic potential of pharmacogenetics and the need for dedicated studies.

D. Autoimmune channelopathies

The autoimmune channelopathies are a group of neurological disorders in which patients develop autoantibodies to ion channels or related functional proteins on neurons or muscle. In these disorders, the antibodies can causes changes of the synaptic function or neuronal excitability by directly inhibiting ion channel. Several antibody-mediated neuromuscular disorders have been already identified: the ligand-gated receptor ion channel-related disorders [myasthenia gravis and myasthenia gravis without antibodies against acetylcholine receptors (AChR), autoimmune autonomic ganglionopathy (AAG), Rasmussen's encephalitis and neuropsychiatric lupus, stiff-person syndrome (SPS)] and the voltage-gated ion channel-related disorders (Lambert–Eaton syndrome, acquired neuromyotonia, Morvan's syndrome, limbic encephalitis). These diseases differ from congenital myasthenic syndromes (CMS) due to gene mutations (Engel, 2007). The mechanism that initiates the production of autoantibodies is unknown. Different studies have hypothesized viral or bacterial infection or complication of systemic malignancy (paraneoplastic) initiated as an immunological response against cancer. Treatment of autoimmune channelopathies encompasses drugs that can be helpful in the restoration of the normal neuronal function (inhibitors of acetylcholinesterase, AChE) or drugs able to modulate the immune response (corticosteroids and immunosuppressive drugs). Other treatments that remove pathogenic antibodies (plasma exchange) or inhibit production of autoantibodies (intravenous immunoglobulin) are clinically helpful. Given the autoimmune nature of these diseases, at the moment there is no drug able to directly modulate channel activity, with exception of 3,4-diaminopyridine, carbamazepine, and propofol. Here, we report a brief description of drugs currently used in therapy suggesting to consider the exhaustive reviews already published in this field (Table 4.1).

1. Ligand-gated receptor autoimmune channelopathies

Antibodies against muscle AChR cause myasthenia gravis (MG). Current therapies using AChE inhibitors, such as pyridostigmine and edrophonium, restore neuromuscular junction transmission by slowing the breakdown of acetylcholine (Drachman, 1994; Vernino, 2007). Other symptomatic agents used in MG include ephedrine, which increases ACh release (Engel, 2007). Immunosuppressive therapy is based on corticosteroids (Pascuzzi *et al.*, 1984; Schneider-Gold *et al.*, 2005), azathioprine, mycophenolate mofetil (MyM), methotrexate, cyclosporine, and tacrolimus, which inhibit the synthesis and proliferation of immune cells as a final point (Buckley and Vincent, 2005; Matsuda and Koyasu, 2000). Glucocorticoids have also beneficial effects against inflammation. Tacrolimus also increases excitation–contraction coupling in skeletal muscle by affecting ryanodine receptor-mediated Ca^{2+} release (Timerman *et al.*, 1993). Interestingly, MG patients with anti-RyR antibodies rapidly respond to treatment, indicating a symptomatic effect on muscle strength in addition to the immunosuppression. Leflunomide (pyrimidine synthesis inhibitor) and pixantrone (topoisomerase II inhibitors) have been proposed in therapy for their efficacy in the suppression of experimental MG in AChR protein-immunized rats (Ubiali *et al.*, 2008; Vidic-Dankovic *et al.*, 1995). New therapies include TNFα antagonists, such as etanercept, since higher level of this proinflammatory cytokine has been found in T cells of MG patients with respect to healthy individuals (Bongioanni *et al.*, 1999). Also Rituximab, a monoclonal antibody binding the CD20 cell surface protein located on B cells (Cragg *et al.*, 2005), is now being explored for use in autoimmune disorders (Zaja *et al.*, 2000). Antisense oligonucleotide (EN101)-blocking AChE synthesis has been suggested in MG therapy after studies showing its beneficial effect on rat experimental model of the disease (Boneva *et al.*, 2006; Soreq and Seidman, 2000)

Table 4.1. Current and Potential Therapy for the Autoimmune Channelopathies

Autoimmune channelopathy	Target of autoantibodies	Clinically used drugs	Mechanism of action
Myasthenia gravis	Muscle AChR	Pyridostigmine, edrophonium	AChE inhibitors
		EN101 (antisense oligo)	AChE suppression
		Ephedrine	ACh release
		Corticosteroids, azathioprine, MyM, cyclosporin, tacrolimus, methotrexate, pixantrone	Immunosuppression (lymphocyte inhibition)
		Etanercept, rituximab	Anti-TNFα–anti-CD20
		Statins?	Inflammation inhibition
MG without AChR antibodies	MuSkinase (AChR clustering)	Corticosteroids	Immunosuppression
Autoimmune autonomic neuropathy	Ganglionic AChR	Pyridostigmine	AChE inhibition
		Corticosteroids	Immunosuppression
		Sildenafil	PDE5 inhibition
Rasmussen's encephalitis	GluR3	Tacrolimus	Immunosuppression of T-cells
Neuropsychiatric lupus erythematosus	NMDA receptor	Costicosteroids, methotrexate, azathioprine, cyclosporine, cyclophosphamide, MyM	Immunosuppression
		NSAID, chloroquine	Anti-inflammatory action
		Rituximab, epratuzumab	Anti-CD-20, anti-CD-22
Stiff-person syndrome	GABA-R-associated protein	Propofol	GABA agonist
		Rituximab	Anti-CD-20
Lambert–Eaton myasthenic syndrome	VGCC	3,4-diaminopyridine	K^+ channel inhibition
		Pyridostigmine	AChE inhibition
		Prednisone, azathioprine, cyclosporine, MyM	Immunosuppression
		Guanidine hydrochloride	Ca^{2+} modulation
Acquired neuromyotonia	VGKC	Corticosteroids, azathioprine, cyclophosphamide	Immunosuppression
		Phenytoin, carbamazepine	Na^+ channel block
		Gabapentin	Ca^{2+} channel block
Morvan's syndrome	VGKC	Immunomodulatory treatments	
Limbic encephalitis	VGKC	Carbamazepine	Na^+ channel block
		Corticosteroids	Immunosuppression
Neuromyelitis optica	AQP-4	Rituximab	Anti-CD-20
		Mitoxantrone	B cells immunosuppression

and in clinical trials (Argov *et al.*, 2007). Although those new therapies are promising, more clinical evidences are needed before to recommend it. Recent advances indicate innovative treatment options for autoimmune MG: gene transfer to modify antigen-presenting cell (APC) targeting AChR-specific T cells (Drachman *et al.*, 2003); complement inhibition (Piddlesden *et al.*, 1996); administration of antibodies that recognize the binding site for the antigen of the T cell receptor (T cell vaccination) (Cohen-Kaminsky and Jambou, 2005). Statins show also promising effects in preclinical models of autoimmunity, as well as in early-stage clinical trials (Vollmer *et al.*, 2004). A significant number of patients with generalized MG, but without AChR antibodies, show the presence of antibodies directed to muscle-specific kinase (MuSK) (Sanders *et al.*, 2003), a receptor tyrosine kinase involved in AChR clustering (Cole *et al.*, 2008). In these patients, difficulties in obtaining clinical remission result in long duration of steroid treatment with higher doses (prednisolone > 40 mg) (Farrugia *et al.*, 2006).

An antibody-mediated disruption of synaptic transmission in autonomic ganglia is responsible for an acquired form of autonomic failure in adults, AAG. Affected individuals typically experience orthostatic hypotension, reduced lacrimation, gastrointestinal disturbance, atonic bladder, and impotence. The basic treatment consists in the symptomatic management of autonomic failure. The AChE inhibitors (pyridostigmine) may improve ganglionic synaptic transmission as well as muscarinic transmission (Singer *et al.*, 2003; Vernino *et al.*, 2008). Sildafenil helps with sexual dysfunction, while fludrocortisone and pyridostigmine might help with orthostatic hypotension (Buckley and Vincent, 2005).

Rasmussen's encephalitis is a severe chronic inflammatory brain disease associated to antibodies against type 3 glutamate receptor (GluR3). It is usually accompanied by nontractable epilepsy (Bien *et al.*, 2005). Immunotherapies may be beneficial and clinical trials using tacrolimus are now recruiting.

In systemic lupus erythematosus, antibodies interacting with the glutamate-gated NMDA receptor cause neuronal damage (De Giorgio *et al.*, 1991). Lupus is treatable symptomatically with corticosteroids and immunosuppressants, such as methotrexate, azathioprine, cyclophosphamide, and cyclosporine. Nonsteroidal anti-inflammatory drugs can be helpful in reducing inflammation and pain. Also, antimalarial drugs such as hydroxychloroquine and chloroquine are effective. In recent years, MyM has been used as an effective medication, particularly when associated with kidney disease. Most recent research indicates benefits of rituximab and epratuzumab (Dörner and Goldenberg, 2007).

Antibodies against a $GABA_A$ receptor-associated protein (GABARAP), responsible for the expression and stability of the GABA receptor (Dalakas, 2008), have been recognized as a new autoantigen in the SPS. Therapy consists in IV immunoglobuline, plasma exchange, and propofol. CD20-specific monoclonal antibodies are in clinical trials (Kazkaz and Isenberg, 2004).

2. Voltage-gated autoimmune channelopathies

Lambert–Eaton myasthenic syndrome (LEMS) is characterized by the presence of antibodies to peripheral nerve P/Q-type CaChs, which cause aggregation and internalization of functional channels. Many patients have an associated small-cell lung cancer (SCLC). First-line treatment forecasts the KCh blocker 3,4-diaminopyridine, but additional therapeutic effect can be obtained if combined with pyridostigmine. If symptomatic treatment is not sufficient, immunosuppressive therapy should be started, usually with a combination of prednisone and azathioprine. Other drugs like cyclosporin and MyM can be used, although their benefit is limited. Guanidine hydrochloride, although highly effective, is less recommended in LEMS therapy because of its severe side reactions (hematologic abnormalities and renal insufficiency).

Acquired neuromyotonia (NMT) is characterized by peripheral nerve hyperexcitability caused by antibodies to nerve voltage-gated KChs (Hart *et al.*, 2002). Antibodies from patients have been shown to cause a reduction in K^+ currents in cultured cells and repetitive firing of APs, similarly to KCh antagonists (Shillito *et al.*, 1995). Symptomatic treatments with membrane-stabilizing drugs, such as phenytoin, carbamazepine, and gabapentin, are often effective and sufficient in milder cases. Also corticosteroids, azathioprine, and cyclophosphamide can be used.

Morvan's syndrome develops peripheral nerve hyperexcitability, autonomic symptoms (excessive sweating, hypersalivation, cardiac instability), and CNS involvement (insomnia, agitation). In these patients, the presence of KCh antibodies has been reported and only some of them respond to immunomodulatory treatments (Liguori *et al.*, 2001).

Limbic encephalitis (LE) is a paraneoplastic immune-mediated condition associated with SCLC, thymoma, breast or testicular tumors, and with antibodies to markers for the associated tumors (Gultekin *et al.*, 2000). In general, these conditions respond poorly to treatment and improvement might be obtained by successful treatment of the tumor (Dalmau *et al.*, 2004). Patients with a nonparaneoplastic form of LE develop KCh antibodies. The response to immunomodulatory therapy is remarkable. Anticonvulsant drugs (carbamazepine) can be beneficial (Harrower *et al.*, 2006). High-dose oral corticosteroid treatment is particularly important in addition to IV Ig or plasma exchange.

3. Other autoimmune channelopathies

Many patients with neuromyelitis optica, a severe inflammatory demyelinating disorder of the optic nerves and spinal cord, produce antibodies against aquaporin-4, a membrane water channel of the CNS (Lennon *et al.*, 2005). Recent evidences show that rituximab and mitoxantrone, antineoplastic agents that target B cells and macrophages, can be helpful (Jarius *et al.*, 2008; Weinstock-Guttman *et al.*, 2006).

E. Insulin-secreting disorders

There are a large group of disorders of the pancreas regarding insulin secretion that requires clinical care and drug treatments such as pancreatic tumors of α- or β-cells (Tricarico *et al.*, 2008a,b). Insulin resistance is a condition in which normal amounts of insulin are inadequate to produce a normal insulin response from fat, muscle, and liver cells causing elevated blood glucose levels. High plasma levels of insulin and glucose due to insulin resistance, also associated with abnormal function of muscle KATP channel subtypes, often lead to metabolic syndrome and type 2 diabetes, including its complications. Other disorders in which KATP channels are involved are congenital hyperinsulinism and neonatal diabetes mellitus. The therapy is based on the use of several drugs and chirurgical treatment. One group of drugs are the sulfonylureas and glinides which are known blockers of the KATP channels largely used as insulin-releasing agents in the treatment of adult diabetes type II; another group of drugs are the pancreatic KATP channel openers which are represented by diazoxide. Other drugs are also CaCh blockers like nifedipine, hormones like insulin, glucagon, and somatostatin which may indirectly affect ion channel activity.

F. Transepithelial transport and vesicular channelopathies

1. Cystic fibrosis

Although over the last two decades our knowledge of the genetic defect and physiopathology of cystic fibrosis (CF) has progressed vastly, there is at the present no curative treatment for this channelopathy. To date, gene therapy has failed to demonstrate a clinical benefit and the current therapy is mainly restricted to an alleviation of symptoms (Proesmans *et al.*, 2008).

CF is a complex multisystem disease (Accurso, 2008; Rowe *et al.*, 2005). The most serious cystic fibrosis symptoms are generally observed in the lungs where the fluid covering the airway epithelia becomes viscous and susceptible to bacterial infection. Other CF symptoms include pancreatic dysfunction, elevated sweat electrolytes, and male infertility. Currently, CF treatment is based on the use of antimicrobials (chinolones, macrolides, lincosamides) for eradicating lung bacteria infection, mucolytics (*N*-acetylcysteine, human recombinant DNAse) for reducing viscosity in the lungs and promoting secretion clearance, as well as steroids and nonsteroidal anti-inflammatory drugs (Proesmans *et al.*, 2008).

Furthermore, inhaled osmotic active substances (hypertonic saline, mannitol) that draw water onto the airway surface are frequently used to restore normal airway epithelia hydration.

Importantly, it must be underlined that together with impaired cAMP-dependent Cl^- secretion, enhanced Na^+ absorption by the apical membrane epithelial Na^+ channel (ENaC) occurs in CF airways. Consequently, a possible therapeutic approach is to tackle these secondary ion transport disturbances by blocking the overactive ENaC other than by activating non-CFTR Cl^- channels. Particularly, amiloride and its long-acting analogues as ENaC blockers (Rodgers and Knox, 1999) and MOLI-1901 as opener of alternative Cl^- channels (Grasemann *et al.*, 2007; Zeitlin *et al.*, 2004) are the drugs used to this aim. Finally, administration of denufosol, a synthetic nucleotide capable of stimulating ATP-activated purinergic receptors (P2Y2 receptor pathways), represents the most promising current strategy (Deterding *et al.*, 2007). Indeed, such a strategy has the advantage of simultaneously activating Ca^{2+}-activated Cl^- channels (CaCC) and inhibiting epithelial Na^+ channels.

Approaches aimed at correcting the basic CF defect still hold promise for curing the disease. The main goal is to improve channel gating (potentiator) or to correct mutated CFTR function (correctors). In this context, attention is directed toward mutation class-specific therapy. This field of research is expanding rapidly, primarily due to the use of the "high-throughput screening" rapidly automated methods for identifying and analyzing potentially active compounds. Due to the rapid changes in this area and the extensive reviews already available in literature, we recommend to consult the CFTR pharmacology elsewhere.

2. Bartter's syndromes

Bartter's syndrome (BS) is a group of closely related hereditary tubulopathies, two of which are due to mutations in genes encoding ion channels as ROMK (type II) and CLC-Kb (type III) (Hebert, 2003). Besides, disruption of the gene encoding barttin, the β-subunit essential for CLC-K channels expression and functionality, leads to type IV BS (Birkenhäger *et al.*, 2001). The associated phenotypes of the various BS types are highly variable and may present either as a typical antenatal variant with or without deafness or as a classic Bartter's syndrome characterized by an onset in infancy or early childhood (Konrad *et al.*, 2000; Naesens *et al.*, 2004). In addition to marked salt wasting, patients have polydipsia, polyuria consequent to hypokalemia, volume contraction, muscle weakness, and growth retardation. The primary symptoms of BS lead to a secondary increase in prostaglandins as a consequence of volume contraction, and many clinical problems are related to elevated prostaglandins (Hebert, 2003).

BS therapy remains empirical (Kleta and Bockenhauer, 2006). In practice, drug and replacement (K^+, Mg^{2+}, and sometimes Na^+) therapies are essentially guided by symptom relief. NSAIDs such as indomethacin, angiotensin-converting enzyme inhibitors, and aldosterone antagonist have been usually used with varying degrees of success and tolerability. However, the use of these drugs to treat these syndromes is not an approved indication. Indeed, one limiting factor in the up-to-day available treatment is the drug-induced risk of progressive renal damage, which can finally lead to chronic renal failure (Unwin and Capasso, 2006).

In the last few years, the pharmacological inhibition and activation of CLC-Ka/barttin and CLC-Kb/barttin has been studied in considerable details, opening the way toward identification of drugs potentially useful for Bartter's syndrome or as diuretics (see Fig. 4.4 and Section III).

3. ENaC-related diseases

Mutations of the epithelial NaCh ENaC, a heteromeric complex of three homologous proteins with unknown stoichiometry, lead to either hereditary hypertension or hypotension. Particularly, mutations in the α-, β-, and γ-subunit of ENaC that lead to loss of function of channel activity are associated with autosomal recessive pseudohypoaldosteronism type 1 (PHA1) (Chang *et al.*, 1996; Gründer *et al.*, 1997). This disorder is characterized by marked hypotension and dehydration of newborns and infants due to an excessive loss of Na^+ and water. The therapy is limited to lifelong Na^+ supplementation and treatment with K^+-binding resins (O'Shaughnessy and Karet, 2004), although small molecule activators of human ENaC heterologously expressed have been more recently described (Lu *et al.*, 2008).

Contrary to the autosomal recessive type I PHA, Liddle syndrome is caused by gain-of-function mutations in the ENaC gene (Hansson *et al.*, 1995; Snyder *et al.*, 1995). A state of pseudohyperaldosteronism is the result of such hyperfunction of ENaC, leading to volume expansion and hypertension on one hand, but also to secondary activation of ROMK and thus increased K^+ excretion at the cortical collecting duct, resulting in hypokalemia. Other than requiring a low-salt diet, these patients respond well to the administration of the diuretic amiloride or triamterene (O'Shaughnessy and Karet, 2004). This clinical observation is understandable in the light of the known competition between these agents and Na^+ at the level of the ENaC-conducting pore.

4. Transient receptor potential (TRP) channel-related disease

TRPM6 is a Mg^{2+}-permeable channel, belonging to the TRP family of cation channels, expressed in the apical membrane of distal convoluted tubule and brush-border membrane of absorptive cells in duodenum (Hoenderop and

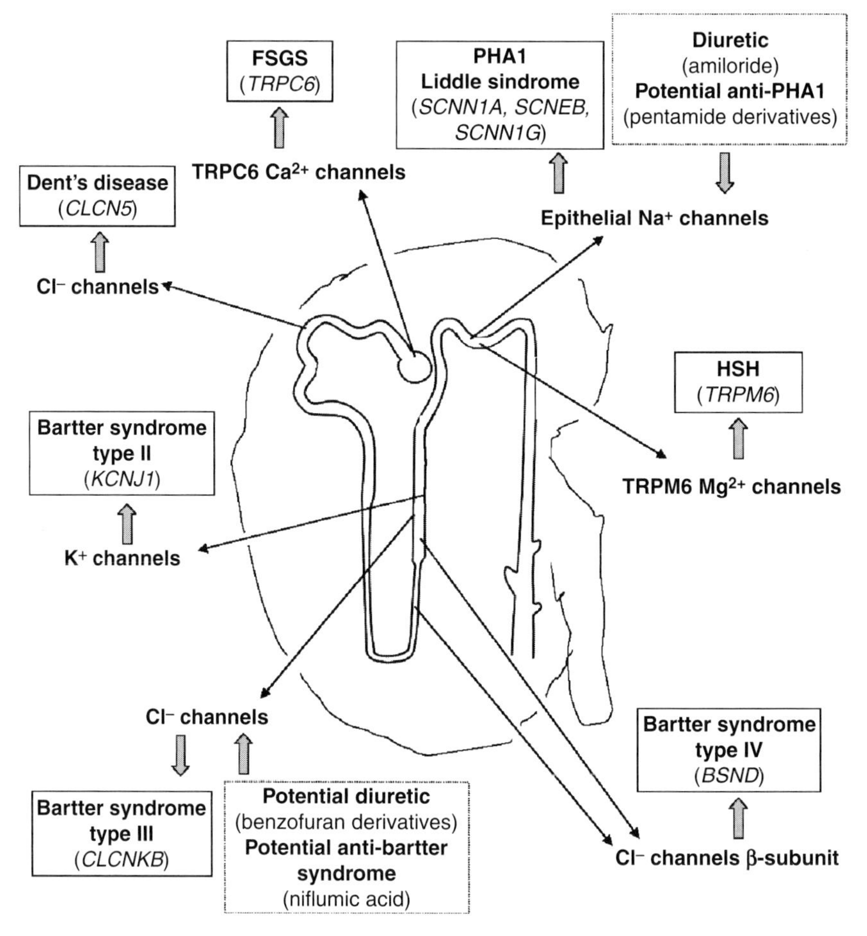

Figure 4.4. Ion channels involved in channelopathies of skeletal muscle and target of drugs. Channels and related genes, classes of drugs with examples, and therapeutic indications are reported. Modified from Ashcroft (2006).

Bindels, 2005). TRPM6 mutations cause hypomagnesemia with secondary hypocalcemia (HSH), an autosomal recessive disorder of intestinal and renal magnesium transport (Schlingmann *et al.*, 2002; Walder *et al.*, 2002). Patients with HSH present with generalized seizures due to profound hypomagnesemia and hypocalcemia during the first months of life and require intravenous Mg^{2+} during convulsive episodes and lifelong high-dose oral Mg^{2+} replacement (Naderi and Reilly, 2008). Heterologously expressed TRPM6 was found to form a Mg^{2+}- and Ca^{2+}-permeable channel, which is regulated by Mg^{2+} and blocked by ruthenium red (Harteneck, 2005). Although the development of TRP-modulating drugs is currently noteworthy ongoing, no TRPM6-specific pharmacological tools are available.

Missense mutations in TRPC6, which cause an increase in Ca^{2+} transient and a gain of function in this channel located on the podocyte cell membrane, are responsible for a form of familial focal segmental glomerulosclerosis (FSGS) (Winn *et al.*, 2005). Typical manifestation of FSGS includes proteinuria, hypertension, renal insufficiency, and eventually kidney failure. The classical therapy in these diseases consists in blocking the renin–angiotensin system by angiotensin-converting enzyme inhibitors or angiotensin-receptor blockers (Mukerji *et al.*, 2007). Although no ligand for TRPC6 is known, there is no doubt that TRPC6 may represent a useful therapeutic target in treating chronic kidney disease. The goal would be to produce a molecule that is highly specific blocker of TRPC6.

5. Dent's disease

In agreement with its role in vesicular acidification, it is recently demonstrated that CLC-5 is an electrogenic Cl^-/H^+ antiporters rather than a channel (Picollo and Pusch, 2005; Scheel *et al.*, 2005). Human mutations in CLCN5 lead to low molecular weight proteinuria, urinary loss of phosphate and Ca^{2+}, and frequently to kidney stones in a syndrome called Dent's disease (Lloyd *et al.*, 1996). Therapy is aimed to reduce stone risk by a long-term control of hypercalciuria by using thiazide diuretics or high citrate diet (Cebotaru *et al.*, 2005; Ludwig *et al.*, 2006). When renal insufficiency occurs, the patients may eventually require kidney transplantation. No CLC-5 ligand is up to date available, although newly identified blockers for other renal CLC Cl^- channels led to hope a pharmacological characterization of CLC-5 feasible in the next future (see Section III).

6. Osteopetrosis

Human osteopetrosis is a rare genetic disorders caused by osteoclast failure, which ranges widely in severity. It relies on loss-of-function mutations of various gene, including the *TCIRG1* gene, encoding for the α_3-subunit of the H^+-ATPase, the *CLCN7* and the *OSTM1* genes, which have closely related function (Del Fattore *et al.*, 2008). Until now, more than 30 *CLCN7* mutations have been identified (Jentsch, 2008). This gene encodes the Cl^- channel coupled to the vacuolar proton pump in the osteoclast. CLC-7 mutations could lead to infantile malignant osteopetrosis (patients showed a complete loss of function of CLC-7 Cl^- channel) or to a milder variant, termed intermediate autosomal recessive osteopetrosis (patients appeared to have some residual function of CLC-7 Cl^- channel).

CLC-7 needs a small integral membrane protein, Ostm1, as auxiliary β-subunit (Lange *et al.*, 2006). Most likely due to a loss of CLC-7 activity, mutations in OSTM1 also underline rare cases of human recessive, malignant infantile osteopetrosis (Chalhoub *et al.*, 2003; Ramìrez *et al.*, 2004).

Osteopetrosis patients show heterogeneous clinical manifestations. Usually, skeletal abnormalities include the loss of bone marrow cavities that are instead filled by bone material, as well as a failure of teeth to erupt (Kornak *et al.*, 2001). Life-threatening symptoms are anemia, pancytopenia, osteomyelitis, and sepsis due to poorly developed bone marrow and impaired medullary hematopoiesis, associated with secondary hematopoiesis in liver and spleen that causes hepatosplenomegaly (Del Fattore *et al.*, 2008). Mutant CLCN7 or OSTM patients show additional manifestation such as lysosomal storage disease in the brain and retina, which is responsible for the degeneration of these nervous tissues (Kasper *et al.*, 2005; Lange *et al.*, 2006). To ameliorate the symptoms, patients are given blood transfusion and are treated for infections. Therapeutic strategy based on administration of calcitriol and IFNγ to reverse phenotype or slow disease progression has been often used but with minimal success. At present, the only curative therapy is hematopoietic stem cells transplantation (Askmyr *et al.*, 2008; Del Fattore *et al.*, 2008). Attempts are ongoing to identify specific CLC-7 ligands for developing potential drug useful in treating osteopetrosis (CLC-7 openers) or osteoporosis (CLC-7 blockers) (see Section III).

III. MOLECULAR PHARMACOLOGY OF ION CHANNELS

A. K^+ channels

Three major families of KChs are defined on the basis of the primary amino acid sequence of the pore-containing unit (α-subunit): voltage-gated KChs containing six or seven transmembrane regions with a single pore, including also KCNQ, hERG, and the Ca^{2+}-activated K^+ channels; inward rectifiers (Kir) containing only two transmembrane regions and a single pore; and two-pore tandem KChs containing four transmembrane segments with two pores. The pore subunits coassemble with auxiliary subunits, affecting their pharmacological responses, trafficking and modulation by second messengers.

1. Voltage-gated K^+ (Kv) channels

The Kv channels are drug targets in several neuromuscular disorders. Several members of related Kv genes were identified in mammals and divided into eight gene families: KCNA (Kv1.1–8), KCNB (Kv2.1–2), KCNC (Kv3.1–4), KCND (Kv4.1–3), KCNF (Kv5.1), KCNG (Kv6.1–4), KCNS (Kv9.1–3), and KCNV

(Kv8.1–3). The Kv1–4 families form homomeric channels with other subunits within their own family or with the electrically silent families (Kv5, Kv6, Kv8, and Kv9). These channels are involved in several neuronal and muscular channelopathies, being also target of drugs. The Kv channels have been investigated by using peptide toxins from animals and plants, such as dendrotoxins, kaliotoxin, hongotoxin, margatoxin, and others, which block the channel pore at picomolar to nanomolar concentrations and serve as tools for the analysis of their structure–function relationships. These toxins block Kv1.1–6 channel subtypes. Although Kv channels were the first to be molecularly characterized, no selective blockers or openers are available. Tetraethylammonium (TEA) and 4-aminopyridine (4-AP) are classic Kv channel blockers, which can discriminate between various channel subtypes. The Kv1.1, Kv3.1–4, and Kv7.2 channels are more sensitive to TEA. Kv1.1–5, Kv1.7, and Kv3.1–2 are inhibited by micromolar concentrations of 4-AP, but millimolar concentrations are needed to block Kv1.6, 1.8, 2.2, 3.3, and Kv4.1–3. Other members such as Kv2.1, Kv3.4, hERG, eag1, and KCNQ channels are insensitive to 4-AP. Several 4-AP analogues have been tested against Kv channels, and the order of potency as Kv inhibitors ranks as follows 3,4-DAP > 4-AP > 3-AP > 2-AP. These drugs cause neuronal firing and release of neurotransmitters. Thus, 4-AP and 3,4-diaminopyridine (3,4-DAP) (25–60 mg/day) are effective in those conditions associated with loss or reduced quantal release of neurotransmitters such as MG, LEMS, and some neuronal channelopathies. The Kv blockers are not effective in neuromyotonia, however, which is characterized by peripheral nerve hyperexcitability that responds to AEDs, including carbamazepine, phenytoin, and pregabalin. Other nonselective Kv channel blockers include linopirdine, a KCNQ channel blocker that evokes quantal ACh release in the CNS; the antiarrhythmic drugs flecainide and bupivacaine (100–250 μmol/L); the CaCh antagonists verapamil, nifedipine, nicardipine, diltiazem (20–200 μmol/L); and riluzole (100–200 μmol/L), a neuroprotective agent used in treating amyotrophic lateral sclerosis. Kv channel openers are not available.

2. KCNQ channels

The KCNQ channels are responsible for the muscarinic currents (M-currents) in neurons and are responsible for repolarizing K^+ currents in cardiac tissues (Miceli *et al.*, 2008). KCNQ channel blockers were developed as cognition enhancers, such as linopirdine and XE-991. Blocking of M-currents underlies the enhancement of transmitter release by these drugs. Linopirdine increases ACh release in rat brain and improves performance in animal models of learning and memory. Although clinical data with linopirdine were inconclusive, several analogues were developed as orally active ACh-releasing agents with potential in

Alzheimer's disease. Heteromers derived from the cardiac KCNQ1/minK are 14- to 18-fold less sensitive to these blockers compared with either KCNQ1 alone or neuronal KCNQ2/3 combination, demonstrating selectivity of these compounds for neurotransmitter release over cardiac function. Compounds inhibiting M-currents selectively are potentially useful for treating cognitive deficits in neurodegenerative diseases and are likely to be forthcoming. These compounds are also promising drugs in the treatment of neuronal channelopathies.

KCNQ channel openers are the newest antiepileptic agents. Retigabine, the desaza-analog of flupirtine (approved in Europe for general nociceptive pain), was originally identified as an anticonvulsant. Retigabine is effective in various epilepsy models and was shown to activate muscarinic currents in various types of cultured neurons. Retigabine acts on all neuronal KCNQ subunits, but not on the cardiac KCNQ1. Analogues of retigabine with more favorable pharmacological profiles are under investigation including the benzanilide derivative ICA-27243, which is a more selective KCNQ opener.

3. Ether-a-go-go-related channels (hERG)

The hERG channels are highly expressed in cardiac myocytes, where they function to restore resting membrane potential following AP generation (Sanguinetti and Tristani-Firouzi, 2006). These channels are blocked by numerous drugs belong to different classes, including antibiotics, antiarrhythmics, antihistaminics, antipsychotics, and others. Investigation on hERG blockers has a great toxicological relevance, since hERG blockers can prolong the QT interval on the ECG, increasing the risk of torsades-de-pointe arrhythmias with fatal events, especially in predisposed individuals (Zünkler, 2006). Every year, new entities are added to the list of hERG blockers, while this acquired form of LQT syndrome has been the single most common cause of drug relabeling or withdrawal of marketing drugs in the last years.

Although the great effort in investigating on possible hERG channel openers, some of these compounds have been described only recently (Zeng *et al.*, 2006). Such compounds might be of interest as antiarrhythmic drugs. RPR260243, PD-118057, and NS1643 are distinct chemical entities that activate hERG channels through a variety of mechanisms. Mallotoxin (MTX) is a natural occurring substance extracted from the tree *Mallotus philippinensis*, which is a potent and unique hERG channel activator. MTX was originally described as an inhibitor for protein kinase C (PKC), Ca^{2+}/calmodulin-dependent protein kinase II and III, and elongation factor-2 kinase, and recently it has been proposed as activator of BK and hERG channels. MTX increases both step and tail hERG current by leftward shifting the voltage dependence of hERG activation and slowing channel deactivation at submicromolar concentrations.

MTX represents a useful tool for the investigation of hERG channel physiology and channelopathies related to LQT syndrome, SQT syndrome, and cardiac arrhythmias.

4. Calcium-activated K^+ (KCa) channels

Three subfamilies of Ca^{2+}-activated K^+ channels can be distinguished. The first one comprises the large conductance channels KCa1.1 encoded by *KCMNA1* (slo1) gene (BK), and the KCa4.1–2 and KCa5.1 channels, which are, however, less Ca^{2+} sensitive. BK channels are widely expressed and are involved in hypertension, coronary artery spasm, urinary incontinence, stroke, psychoses, and several neurological disorders including epilepsy and schizophrenia. The second subfamily consists of the intermediate conductance KCa3.1 channel encoded by the *KCNN4* gene (IK), expressed in erythrocytes and thymocytes, where it plays a role in immunostimulation. This channel is also expressed in several cancer cell lines being involved in cell proliferation. The third of the subfamilies includes the small conductance KCa2.1–3 channels encoded by the *KCNN1–3* genes (*hSK1–3*), found in a variety of cells including sympathetic neurons, intestinal smooth muscle, bladder smooth muscle, hepatocytes, and brown adipocytes. In many excitable cells, the SK channels are responsible for the slow afterhyperpolarization that follows AP. Calmodulin is associated with the SK α-subunit and is necessary for Ca^{2+} binding and gating.

The BK channel openers stabilize the cell by increasing K^+ efflux in response to intracellular Ca^{2+} rise, leading to hyperpolarization and thus decrease of cell excitability. The BK channel has been more appealing as a therapeutic target than KATP channels because of the lower expression of BK channels in the heart. Furthermore, extensive K^+ efflux and a late channel opening of BK channels circumvent the adverse cardiac effects associated with KATP channel openers. The different subunit compositions of BK channels in various tissues open the possibility of finding tissue-selective BK openers. Indeed, the skeletal muscle BK is composed of the α-subunit alone, the vascular BK is composed of $\alpha + \beta_1$, and the neuronal types of $\alpha + \beta_4$ or $\alpha + \beta_3$. These channels show different responses to modulators. Peptide toxins can discriminate between peripheral BK formed by α-subunit alone, which is sensitive to charybdotoxin (ChTX), and neuronal BK channels formed by $\alpha + \beta_4$-subunits, which is resistant to ChTX but sensitive to iberiotoxin (IbTX). Slotoxin, from a scorpion venom, selectively blocks α-subunit of mammalian BK channels ($K_d = 1.5$ nmol/L) and can distinguish among α, $\alpha + \beta_1$, and $\alpha + \beta_4$ more efficiently than can IbTX. The BK openers comprise a large series of synthetic benzimidazolone derivatives such as NS004 and NS1619, biaryl amines, biarylureas, pyridyl amines, 3-aryloxindoles, benzopyrans, dihydropyridines, and natural modulators such as dihydrosoyasaponin-1 (DHS-1) and flavonoids. Both NS004 and NS1619 are known as α-subunit-selective BK openers. The 3-fluoro aryloxindole analogue BMS-204352 is a neuroprotective in animal models without affecting heart rate and mean arterial pressure in conscious dogs. In hippocampal slices, it was able to reduce glutamate release. BMS-204352 was well tolerated in phase I and II clinical trials, but failed to show efficacy against placebo as an antistroke agent in phase III. Recently, BMS-204352 showed dose-related anxiolytic efficacy due to activation of KCNQ2–5 channels, and the *R*-enantiomer also activates $GABA_A$ receptor. It is likely that this drug would be of benefit in other channelopathies. Other than benzimidazolone derivatives, a wide structural diversity of drugs showing BK activation properties has emerged (Fig. 4.1). ACTZ, DCP, and some other related drugs increase K^+ currents in membrane patches isolated from rat muscle fibers in the micromolar concentration range by interacting with BK channels and their effects are structure related (Tricarico *et al.*, 2004). The order of potency as BK channel openers is ACTZ $>$ bendroflumethiazide $>$ ethoxzolamide $>$ DCP. Their action as BK channel openers is not correlated with the inhibition of the carbonic anhydrase enzymes, since bendroflumethiazide is not an inhibitor of these enzymes. ACTZ and related drugs are effective in preventing the insulin-induced paralysis and in restoring the serum K^+ levels in K^+-depleted rats, thus explaining their efficacy in hypoPP (Tricarico *et al.*, 2006a). We have identified a BK channel in slow-twitch muscle that is resistant to ACTZ, leading to the idea that different varieties of BK channels may be expressed in the various muscles, which differently respond to drugs. This may contribute to loss of efficacy of ACTZ in some hypoPP patients. Various drugs such as niflumic, flufenamic, and mefenamic acids, as well as 17-β estradiol, activate BK channels in a nonselective manner. Channel activation by 17-β estradiol could contribute to its nongenomic effect on the vasculature, consisting in acute vasorelaxation. Tamoxifen, an estrogen receptor antagonist with mixed estrogenic properties, activates BK channels at therapeutic concentrations and blocks other ion channels such as Kv channels; this may explain the tamoxifen-induced QT prolongation and arrhythmias. An additive mechanism explaining the neuroprotection by BK openers may be the activation of the mitochondrial BK channels, which couple the intracellular Ca^{2+} levels to the electrical activity of the mitochondria.

The BK channel blockers may have a role in those conditions associated with abnormal BK channel activity. Gain-of-function mutations of the *KCMNA1* gene encoding the slo1 BK α-subunit are linked to generalized epilepsy with dyskinesia. Abnormal function of the BK channels present in the presynaptic terminals affects the release of inhibitory neurotransmitters. Neuronal BK channels composed of a β_3-subunit variant (β_3b-V4) fail to terminate AP and hence contribute to neuronal excitability observed in idiopathic epilepsy. Currently, no selective blockers are available for clinical use. In this respect, the antiepileptic effects of carbonic anhydrase inhibitors including ACTZ, zonisamide, and sulthiame may be related to their ability to lower intracellular pH through inhibition of neuronal CA enzymes with reduction of the neuronal firing. Lowering of intracellular pH is known to inhibit several ion channels, including neuronal BK channels. Other unselective BK blockers are verapamil and gallopamil, which produce a flickering block of vascular BK channels, the anesthetic ketamine that inhibits BK channels by an indirect mechanism and the antifungal clotrimazole, which increases hormonal secretion and neuronal excitability by inhibiting BK channels.

Blockers of IK channels may be of therapeutic interest for immunosuppressive therapy, through modulation of thymocytes and erythrocytes function. Nonselective IK blockers such as clotrimazole have shown antiproliferative effects on lymphocytes and cancer cell lines. Clinical evaluation of these drugs is underway. Openers of IK channels may be beneficial in hypertension, cystic fibrosis, and peripheral vascular disease. Although not highly specific, 1-ethyl-2-benzimidazolinone (1-EBIO) and clinically used benzoxazoles are described as pharmacological activators of the IK channel.

SK channel blockers have been suggested for the treatment of myotonic muscular dystrophy, in which an abnormal activation of this channel has been found. They are also proposed in the treatment of gastrointestinal dysmotility, memory disorders, epilepsy, narcolepsy, and alcohol intoxication. Three classes of SK blockers are known: peptide toxins such as apamin and leiurotoxin I (scyllatoxin), bis-quinolinium blockers, and neuromuscular blockers such as tubocurarine. Openers of SK channels may be important in diseases involving loss of synaptic plasticity, including age-related loss of memory and learning in Alzheimer's disease.

5. Inward rectifier and ATP-sensitive K^+ channels

Since the initial cloning of the first inward rectifiers Kir1.1 (ROMK1) and Kir2.1 (IRK1), new members of this family have been identified, including the G protein-coupled Kir3 and the ATP-sensitive Kir6.2. These channels play important roles in many organs including brain, heart, kidney, endocrine cells, ear, and retina. Seven Kir subfamilies are known: Kir1.1 (KCNJ1), Kir2.1–4 (KCNJ2,4,12,14), Kir3.1–4 (KCNJ6,5,9,3), Kir4.1–2 (KCNJ10,15), Kir5.1 (KCNJ16), Kir6.1–2 (KCNJ8,11), and Kir7.1 (KCNJ13). Kir2.1 is expressed in heart, skeletal muscle, and several brain areas, coassembling with other Kir2 to form functional channels. ATP opens Kir2 channels, possibly through phosphorylation by PKA or PIP2. Loss-of-function mutations of Kir2.1 are linked to Andersen–Tawil syndrome. KATP channels are octameric complexes of Kir6.1–2 and the sulfonylurea receptor subunits (SUR1, SUR2A, and SUR2B) with 1:1 stoichiometry. These channels are metabolically regulated, coupling cell energy status with membrane potential. KATP channels are involved in several physiopathological processes involving glucose metabolism and heart and skeletal muscle contractility (Ashcroft, 2006; Nichols, 2006; Tricarico *et al.*, 2006b). In skeletal muscle, a reduced activity of KATP channels in human hypoPP patients carrying the R528H mutation of dihydropyridine receptor was found (Tricarico *et al.*, 1999). Similarly, in K^+-depleted rats, a reduced expression or activity of the Kir6.2/SUR2A subunits was observed in fast twitch muscles, suggesting a contribution of this channel to the hypoPP phenotype (Tricarico *et al.*, 2008b). Other Kir channels play a role in epithelial transport of K^+ ions, including the ROMK1 channel that is associated with the renal Bartter's syndrome characterized by hypokalemia. Kir3.1 (GIRK1) forms heteromeric channels with other members of the family, which are activated by G protein (G-α-subunit) and by PIP2.

Knowledge of the tissue-selective expression of various SUR subunits (SUR1, SUR2A–B) and Kir6.1–2 constituting KATP channels has made possible the search for tissue-selective openers. The SUR subunits of KATP channels are indeed the receptors for antagonists of the pancreatic channels, such as the sulfonylureas (glibenclamide and tolbutamide) and glinides (nateglinide and repaglinide) developed as antidiabetic drugs, and for the KCO–KATP developed as antiangina, antihypertensive, and antihypoglycemic drugs. First-generation KCO–KATP include benzopyrans (cromakalim), cyanoguanidines (pinacidil), thioformamides (aprikalim), thiadiazines (diazoxide), and pyridyl nitrates (nicorandil, minoxidil). Diazoxide activates the pancreatic Kir6.2/SUR1 channel. The main cardiac and skeletal muscle Kir6.2/SUR2A complex is activated by cromakalim and pinacidil, while cromakalim, diazoxide, nicorandil, minoxidil, and pinacidil activate the vascular Kir6.1/SUR2B complex (Jahangir and Terzic, 2005). However, the first-generation KCO–KATP have limited use, in that their lack of tissue selectivity contributes to side effects. Second-generation KCOs include cyclobutenediones, 2*H*-1,4-benzoxazine derivatives, dihydropyridine-related structures (ZM244085), and tertiary carbinols (ZD6169), all showing enhanced potency and tissue selectivity. KCO–KATP are effective in neuromuscular disorders in which they repolarize muscle fibers when the ATP/ADP ratio increases, reducing Ca^{2+} influx and electrical activity. Diazoxide is a mito-KATP opener, causing swelling, stimulation of respiration, inhibition of the MPTP pore, and Ca^{2+}-overload reduction in mitochondria, thereby contributing to neuronal and cardiovascular protection (Wang *et al.*, 2006; Wu *et al.*, 2006). However, first-generation KCO–KATP are not skeletal muscle selective. New benzopyran derivatives, such as 2*H*-1,4-benzoxazine analogues, have been synthesized and tested on the muscular KATP channels (Rolland *et al.*, 2006; Tricarico *et al.*, 2003, 2008b). These compounds open native KATP channels in the nanomolar concentration range in the presence of ATP, while inhibiting it at micromolar concentrations, showing a peculiar bell-shaped dose–response curve. In the absence of nucleotides, only inhibition is observed, indicating the presence of the inhibitory and activatory sites for these compounds on KATP channel subunits. Structure–activity relationship investigations in parallel with molecular modeling identified the molecular determinants for the activating/blocking actions of these compounds acting as agonist/antagonists or as pure agonists. The openers are 100-fold more potent and effective than first- and second-generation KCO against the muscle KATP channels; the potency of the blockers is comparable with that of the sulfonylureas. The observed structural similarities of the 2*H*-1,4-benzoxazine derivatives with the ATP and ADP molecules help to explain their interaction with the nucleotide-binding sites on the KATP channel subunits, such as the recognition sites for ATP and ADP on the Kir6.2 subunit and on the SUR subunits (nucleotide-binding folds: NBD1 and NBD2) (Nichols, 2006). Tissue-selective experiments are ongoing to identify the molecular determinants responsible for the blocking/activating actions of these drugs on pancreatic KATP channels.

The KATP channel blockers, such as sulfonylureas and glinides, are used in the insulin-secreting disorders and are effective in neuromuscular channelopathies associated with neonatal diabetes type II (especially sulfonylureas). It is emerging that their action is not selective against the pancreatic β-cells and are slowly reversible leading to several side effects. These drugs bind to mitochondrial SUR1 in different tissues leading to apoptosis and atrophy. Blockers with a different mechanism of action are under investigation. These belong to the class of the 2*H*-1,4-benzoxazine derivatives which target the Kir6.2 subunit (Tricarico *et al.*, 2008b). Such compound should guarantee a fast block, rapidly reversible, of the KATP channel pore without affecting SURs to present a better pharmacological profile than classical KATP blockers. On the other hand, SUR1-selective blockers may have a promising role in CNS ischemic disorder such as the spinal cord injury (Simard *et al.*, 2008).

6. Two-pore K^+ ion channels (K2P)

The K2P channels are responsible for the background K^+ conductance in the cells at rest. Fifteen mammalian genes belong to the KCNK family encoding the K2P channels, including *TASK1–3*, *TREK1–2*, *TRAAK*, *TWIK1–2*, *TALK1–2*, and others. These are controlled by several stimuli, including oxygen tension, pH, lipids, mechanical stretch, neurotransmitters, and G protein-coupled receptors. These channels are also the targets for volatile and local anesthetics. TASK is involved in chemoreception; its inhibition by extracellular protons or hypoxia depolarizes the cells and starts the firing of respiratory motor neurons with increased frequency in the respiratory reflexes. TREK channels are expressed in neurons involved in thermoregulation. These are modulated by lipids and fatty acids. Volatile anesthetics open TREK channels, whereas local anesthetics block these channels. The fact that the neuroprotective agent riluzole is an activator of TREK raises the question as to whether it can be a drug target for neuroprotection.

B. Voltage-gated Ca^{2+} channels (Cav)

Ten different genes encode different α-subunits composing the voltage-gated CaChs. The Cav1.1–4 (α_1S, α_1C, α_1D, and α_1F) mediate L-type Ca^{2+} currents; Cav2.1–3 (α_1A, α_1B, and α_1E) mediate P/Q-type, N-type, and R-type Ca^{2+} currents, respectively; and Cav3.1–3 (α_1G, α_1H, and α_1I) mediate T-type currents. The α-subunits coassemble with β-, δ-, and γ-subunits to form functional channels in different tissues. Expression of these channels is tissue specific and a number of them have been linked to inherited channelopathies.

The Cav1.1–1.4 subunits have similar pharmacology and are targets of phenylalkylamines, dihydropyridines, and benzothiazepines, which are largely used as antihypertensive and antiarrhythmic drugs. The use of CaCh antagonists in the treatment of hypoPP patients was not successful. In contrast, Cav2.1–3 channels are insensitive to classical CaCh antagonists but are specifically blocked by high-affinity peptide toxins. Selective antagonists would be of benefit in EA2–6 and migraine disorders. Blockers of Cav2.2 channels have strong anti-inflammatory and analgesic effects comparable or superior to opiates. This channel is indeed the target of cannabinoids, opioids, neuropeptide Y, and substance P. Intrathecal administration of ziconotide, a synthetic analogue of the ω-conotoxin MVIIA, is effective in patients not responsive to opiates. Additionally, this compound is free of dependence as compared to opiates. Nonselective blockers of these channels are mibefradil, piperazines, gabapentin, and volatile anesthetics. Cav3 channel antagonists are not related to drugs targeting other CaChs. Mibefradil is fairly selective for T-type versus L-type Ca^{2+} currents but was withdrawn from the market for its low pharmacotoxicological profile. The peptide kurtoxin inhibits activation gating of Cav3.1 and Cav3.2 channels. Other nonselective blockers are penfluridol, pimozide, amiloride, and phenytoin. More specific and high-affinity blockers of Cav3 channels would be useful for therapy. These channels are the main target of ethosuximide, widely used against absence seizures. Other AEDs with multiple modes of action such as zonisamide and valproate can inhibit T-type CaChs.

C. Sodium channels

1. Voltage-gated Na^+ channels

Voltage-gated NaChs allow the initiation and propagation of APs in most excitable cells. The human genome contains nine genes encoding the main α-subunit of voltage-gated NaChs (*SCN1A*, *SCN2A*, etc.) and at least four genes encoding auxiliary β-subunits (*SCN1B* to *SCN4B*). The expression of α-subunits is tissue specific, with Nav1.4 expressed exclusively in adult skeletal muscle, Nav1.5 expressed mainly in the heart, Nav1.7, Nav1.8, and Nav1.9

expressed in peripheral neurons, Nav1.1, Nav1.2, and Nav1.3 expressed in brain, and Nav1.6 expressed in brain and peripheral neurons. Amino acid homology between α-subunits ranges from 60% to 95%. The α-subunits are transmembrane proteins of about 2000 amino acid residues, constituted of four homologous domains, each composed of six α-helical transmembrane segments. The fourth segment of each domain is rich in charged residues, arginine and lysine, which serves as voltage sensors. The four domains arrange in the membrane to form a Na^+-selective pore. A single α-subunit may associate with one or more β-subunits, which comprise a single transmembrane segment with a short intracellular C-terminus and a longer extracellular N-terminus containing IgG loop motif. The β-subunits modulate α-subunit expression levels, cell surface localization, association with other proteins, as well as voltage dependence and kinetics of the channel. The α-subunit contains at least eight distinct high-affinity binding sites for natural toxins, which affect ion permeation and gating resulting in either inhibition or enhancement of Na^+ currents. For instance, NaChs can be divided into two categories, based on their sensitivity to tetrodotoxin (TTX), a paralytic poison from some puffer fishes. The TTX-sensitive channels are blocked by nanomolar concentrations of TTX, whereas block of TTX-resistant channels requires micromolar TTX concentrations. Thus, natural toxins have provided important tools for channel subtype identification, localization, and purification, definition of the channel structure/activity relationship, and perfecting of binding assays that are in use in the pharmaceutical industry.

Low molecular weight compounds that block NaChs are used in therapy for their ability to slow neurotransmission in a large spectrum of membrane excitability disorders, including cardiac arrhythmias, epilepsies, myotonias, and chronic pain (Clare *et al.*, 2000; Conte Camerino *et al.*, 2007). They are also intensively studied for neuroprotection. The local anesthetics (LAs) are chemically related NaCh blockers, which were first developed as alternatives to cocaine to obtain anesthesia and analgesia. Their cardioprotective effect was discovered later, leading to the birth of class I antiarrhythmics. Most clinical LAs present a tertiary amine associated with a hydrophobic aromatic ring through an amide or ester link. Experimental data suggest that the two ends of the drugs may interact with channel pore-lining amino acids through hydrophobic or π-cation interactions (Ragsdale *et al.*, 1996). Antiepileptic NaCh blockers constitute a more heterogeneous chemical class, but many bind to a molecular receptor matching at least partially that of the LAs (Ragsdale *et al.*, 1996).

Because the LAs bind in the inner cavity of the NaCh pore, both lipophilicity and pKa of the drugs are important determinants of channel block (Hille, 2001). At physiological pH, LAs equilibrate between a liposoluble neutral form, that may reach or leave the receptor across the plasma membrane lipid phase even if the channel is closed, and a protonated form that needs channel opening to enter the pore and inhibit Na^+ currents in a use-dependent manner. In addition, according to the modulated receptor hypothesis, the pore changes conformation during channel activity, thereby modifying the drug-binding affinity as a function of channel state: the binding affinity of LA is far greater for open and/or inactivated NaChs than for closed channels. A meticulous alanine-scanning mutagenesis study has individuated the amino acids involved in high-affinity binding to the inactivated channel within the sixth segments of domains I, II, and IV, with domain IV playing the dominant role (Yarov-Yarovoy *et al.*, 2002). On the other hand, measuring the effects of subtle chemical modification of drugs have increased our knowledge about drug requirements for optimal channel block (De Luca *et al.*, 2000, 2003; Desaphy *et al.*, 1999, 2003). Combining all these data with a channel structural picture derived by analogy from X-ray crystal structure of the KcsA KCh allowed molecular modeling of drug–channel interactions (Lipkind and Fozzard, 2005).

The state-dependent affinity of drugs for NaChs has two important implications for therapy. First, the drugs will block NaChs in a voltage-dependent manner because the transition of the channel from the low-affinity closed state to the high-affinity open/inactivated states depends on membrane voltage. Thus, the block will be greater in neurons with respect to skeletal muscle fibers or in an ischemic area with respect to the healthy tissue. Second, channel block will increase with the frequency of stimulation, because the channel will spend more time in the open/inactivated states and the drug will have less time to dissociate from the channel between two APs. This frequency-dependent (or use-dependent) block is fundamental for the clinical use of LAs and AEDs, allowing their selective action on hyperexcited tissues such as the myotonic muscle or epileptogenic neurons. Indeed, it is generally thought that the greater affinity of current drugs for a specific NaCh subtype depends more on differences in channel gating that secondarily affect receptor accessibility than on subtle differences in the receptor site. Drugs selective for a NaCh subtype are currently lacking, although recent studies may have overcome this difficult issue. Indeed, selective blockers of the Nav1.8 channel have been described, which provide a new powerful tool to discriminate the role of this channel in pain sensation (Jarvis *et al.*, 2007; Kort *et al.*, 2008). The results suggested that selective pharmacological block of Nav1.8 channels *in vivo* produces significant analgesia in animal models of neuropathic and inflammatory pain. Clinical data are warranted to validate the translation from rodents to humans. Because of its obvious role in neuropathic pain, as evidenced by *SCN9A* mutations causing PE and PEPD, the Nav1.7 channel subtype concentrates also much interest in view of developing efficient and safer pain relievers (Priest and Kaczorowski, 2007). Ideally, a similar strategy would allow us to identify safer antimyotonic and antiarrhythmic drugs by binding selectively Nav1.4 and Nav1.5 channels, respectively.

In addition to classical NaCh blockers, a number of drugs acting primarily on other targets have been shown to block NaChs, which likely contribute to their therapeutic or toxic action. For instance, the relevance of NaCh blockade in the analgesic action of TCA is today widely acknowledged. Other such drugs include the histamine receptor antagonist diphenhydramine (Kuo *et al.*, 2000), β-adrenoceptor agonists/antagonists (Desaphy *et al.*, 2003), and the selective serotonin reuptake inhibitor fluoxetine (Lenkey *et al.*, 2006). Recently, we found that the anticholinergic agent orphenadrine blocks Nav1.7 channels, which may explain its analgesic action, while blockade of cardiac and brain NaCh subtypes may contribute to side effects (Dipalma, Desaphy, and Conte Camerino, submitted for publication).

Among NaCh blockers, a special attention should be paid to ranolazine, which was recently approved for treatment of chronic stable angina (Chaitman, 2006). Compared with previous drugs, ranolazine can reduce the effects of ischemia without altering hemodynamic parameters. More recent evidences suggest that ranolazine inhibits selectively the long-lasting Na^+ current in cardiac cells, which in turn reduces the guilty Ca^{2+} overload in ischemic myocyte. Its peculiar mechanism of action on NaChs rends ranolazine a promising antiarrhythmic drug, a hypothesis that merits more experimental and clinical investigations (Makielski and Valdivia, 2006).

Regarding the channelopathies due to mutations in NaCh genes, there is the possibility that the mutations may modify sensitivity of the channel to drugs, either by altering the binding site or by altering channel gating that secondarily alters drug effect. For instance, the fact that mexiletine is useful to many patients suffering from either ClCh or NaCh myotonia suggests that mexiletine treatment is symptomatic: The drug, by blocking NaChs, counteracts the hyperexcitability induced by every genetic defect. However, many NaCh mutations themselves can modify the sensitivity of the channel to mexiletine (Desaphy *et al.*, 2001; Takahashi and Cannon, 2001). We proposed that the voltage dependence of channel availability may be considered as a general index of mutant channel responsiveness to drug (Desaphy *et al.*, 2001), and a similar correlation has been recently proposed for *SCN5A* mutations causing LQT syndrome (Ruan *et al.*, 2007). The mutations that shift this voltage dependency toward more negative potentials increase mexiletine sensitivity, and the heterozygous patients may respond better to therapy owing to the preferential block of mutants with respect to wild-type (WT) channels. On the contrary, mutations that shift positively the availability voltage dependence reduce drug sensitivity, and patients may benefit from a drug having a more specific action on the mutant channel. With the aim of defining a pharmacogenetic strategy to better address treatment in individual myotonic patients, we showed that flecainide may exhibit greater benefit to those patients carrying mutations with a positively shifted availability voltage dependence (Desaphy *et al.*, 2004). Altered response to LAs and AEDs has also been reported for mutations responsible for epilepsy and neuropathic pain (Lucas *et al.*, 2005; Sheets *et al.*, 2007). Thus, growing evidences suggest that the biophysical defect induced by each specific mutation could serve as a prediction index of clinical efficacy of NaCh blockers in the corresponding mutation carriers.

The other aspect of the therapy of Na^+ channelopathies regards those diseases caused by loss of function/expression of NaChs, including hypoPP type 2, severe myoclonic epilepsy of infancy, and some cardiac arrhythmias. In these disorders, drugs able to promote Na^+ currents may be useful to restore a normal excitability to the concerned tissue. Pharmacological chaperones may constitute a promising option. These are low molecular weight compounds that bind selectively to intracellularly retained proteins and promote their proper folding and membrane targeting. For example, a number of genetic defects associated with Brugada syndrome impair NaCh trafficking resulting in haploinsufficiency. It has been shown that NaCh ligands such as mexiletine can rescue mutant channels to the surface membrane (Bezzina and Tan, 2002). Alternatively, such agents could increase the expression of WT channels, which may counteract the haploinsufficiency due to retention of the allelic channel mutant (Desaphy, J. F., and Conte Camerino, D., submitted for publication). However, a number of issues must be addressed before performing clinical trials. For instance, drugs may be able to achieve channel rescue without block, to promote normal gating behavior of the rescued mutant channels, and to not generate hyperexcitability by excess in channel expression.

In summary, compounds targeting NaChs have a remarkably extended field of application. However, currently available drugs display a relatively low selectivity profile, acting quite equally on the various channel subtypes. At present, the sole guaranty for a safety use is provided by the voltage- and use-dependent characteristics of NaCh blockade. In the last 50 years, the discovery of Na^+ channelopathies has underscored the specific role of NaCh subtypes in pathological processes. In parallel, molecular modeling of the drug-binding site has increased our understanding of the complex drug–channel interaction. Such information would help to develop drugs more selective for each channel subtypes/mutants in view of defining more efficient and safer pharmacotherapy.

2. ENaC channels

The diuretic amiloride inhibits transepithelial Na^+ transport across reabsorptive epithelia by blocking apical membrane ENaC function (Benos, 1982). Structure–activity relationship studies allowed us to identify the more potent and specific derivatives phemanil and benzamil exhibiting IC_{50} values $\sim$10 nM (Hirsh *et al.*, 2006).

Conversely, basic research on ENaC activators resulted more hard to carry out. Indeed, molecules reported to increase ENaC Na^+ transport, such as glybenclamide, CPT-cAMP, capsazepine, and icilin, all suffer from low efficacy, low specificity, and/or lack of effect on wild-type channels. However, the first compound S3969 (a pentanamide-based structure) was recently described, which fully and reversibly activates hENaC in heterologous cells (Lu *et al.*, 2008). By increasing the open probability of ENaC channels through interactions requiring the extracellular domain of β-subunit, S3969 exhibits high efficacy (600–700% hENaC activation at 30 μM), as well as potency on wild-type ENaC. Small molecule activators of ENaC may find application in alleviating human diseases, including PHA1, hypotension, and neonatal respiratory distress syndrome, when improved Na^+ flux across epithelial membrane is clinically desirable.

D. CLC chloride channels and transporters

Chloride transporting proteins play fundamental roles in many tissues in the plasma membrane as well as in intracellular membrane. CLC proteins form a gene family that comprises nine members in mammals, at least four of which are involved in human genetic diseases. Particularly, CLC-1, CLC-2, CLC-Ka, and CLC-Kb are Cl^- channels that fulfill their functional roles (such as stabilization of the membrane potential, transepithelial salt transport, and ion homeostasis) in the plasma membrane. The remaining CLC proteins are predominantly expressed in intracellular organelles like endosomes and lysosomes, where together with the V-type H^+-ATPase they are important for luminal acidification (Jentsch, 2008).

Importantly, beyond the inherited diseases of CLC channels/transporters, accumulating data evidence their pivotal roles in physiological events such as neuron and muscle excitability, blood-pressure regulation and bone resorption, providing a solid rationale to consider these proteins potential "druggable" targets in a wide spectrum of nongenetic diseases.

1. CLC-1 channels

The CLC-1 channel is typically expressed in skeletal muscle tissue, where it stabilizes resting membrane potential and helps to repolarize sarcolemma after APs (Jentsch, 2008). This channel sustains the large Cl^- conductance (gCl) of fast muscles (Aromataris *et al.*, 1999; Chen *et al.*, 1997; Pierno *et al.*, 1999; Steinmeyer *et al.*, 1991); indeed, a reduction of gCl due to CLC-1 inhibition decreases the amount of inward current required to depolarize the membrane leading to abnormal hyperexcitability. The importance of CLC-1 function is demonstrated by the observation that loss-of-function mutations of *CLCN1* gene cause myotonia congenita, a Cl^- channelopathy characterized by delayed relaxation (Pusch, 2002). Resting gCl strictly depends on CLC-1 Cl^- channel expression and regulation. A Ca^{2+}- and phospholipid-dependent PKC controls CLC-1 channel activity; indeed, activation of PKC closes the channel (Rosenbohm *et al.*, 1999) and reduces gCl (Bryant and Conte Camerino, 1991). This means that gCl can be modified by Cl^- channel agonists and antagonists but can also be indirectly affected by PKC ligands.

The drugs of interest for CLC-1 channelopathies should be able to increase Cl^- current at the aim to counteract hyperexcitability of myotonia. However, there are very little compounds that respond to this request. The *R*(+)-isomer of 2-(*p*-chlorophenoxy)propionic acid (CPP) was able to increase gCl at concentrations as low as 1–5 μM, making it a promising compound in therapy (Conte-Camerino *et al.*, 1988; De Luca *et al.*, 1992). However, the opener activity of R-CPP was not observed in CLC-1 expressed in heterologous system (Aromataris *et al.*, 1999; Pusch *et al.*, 2000), suggesting an indirect action through a muscle-specific component. Only recently, lubiprostone, a bicyclic fatty acid derivative, approved for the treatment of idiopathic chronic constipation, has been described to activate CLC-2 channels without activation of prostaglandin receptors (Lacy and Levy, 2007). Studies are warranted to verify whether such compound may serve as lead compound for future CLC openers. ACTZ used as an alternative drug in myotonia, has been shown to increase CLC-1 current in a mammalian cell line (Eguchi *et al.*, 2006), most probably through intracellular acidification. We have also observed that ACTZ was able to improve function in a myotonia congenita Cl^- channel mutant. Thus, the indirect activation of wild-type and mutant CLC-1 channels by ACTZ might contribute to its therapeutic effect in muscle disorders (Desaphy *et al.*, 2007). Also, taurine increases either gCl or Cl^- currents sustained by human CLC-1 heterologously expressed in Xenopus oocytes. The amino acid shifts the Cl^- channel activation toward more negative potentials sustaining its effectiveness in myotonic states (Conte Camerino *et al.*, 1989, 2004). Inhibitors of PKC, such as staurosporine or chelerythrine, has been demonstrated to affect Cl^- channel (Rosenbohm *et al.*, 1999) and increase gCl (De Luca *et al.*, 1994; Pierno *et al.*, 2003). Some of them already used in therapy to treat cancer (i.e., enzastaurin) might be tested for ameliorating Cl^- channel loss of function.

Compounds shown to block gCl have been characterized in the past. Anthracene-9-carboxylic acid (9-AC) reproduces, in skeletal muscle of healthy animals, the reduction of gCl and related hyperexcitability found in skeletal muscle of myotonic goats (Bryant and Morales-Aguilera, 1971). Also, clofibrate or its *in vivo* metabolite CPP induces a myotonic-like state in rodents and CPP is effective in blocking gCl in a dose-dependent manner when applied *in vitro*, with an IC_{50} of 15 μM (Conte Camerino *et al.*, 1984; De Luca *et al.*, 1992). Following research has been encouraged toward structure–activity studies by using a considerable number of clofibric acid derivatives at the aim to find more specific compounds with therapeutic usefulness. These studies have been conducted by comparing their effects on gCl measured in native skeletal muscle fibers with those obtained on hCLC-1 channel expressed in cultured cells. The S-isomer of CPP is the most active compound. It inhibits CLC-0 and CLC-1 and with less potency CLC-2 (Conte-Camerino *et al.*, 1988; Liantonio *et al.*, 2002; Pusch *et al.*, 2000). The action of this blocker is voltage dependent by reducing currents at negative voltages (Aromataris *et al.*, 1999) and indicating higher affinity to the closed channel (Pusch *et al.*, 2001). Mutagenesis studies have shown that *p*-chlorophenoxyacetic acid (CPA) exerts its blocking activity by binding the channel pore (Accardi and Pusch, 2003). The obtained results allowed us to make clear the molecular requisites for modulating gCl and for the interaction with the binding site on muscle CLC-1. An "ideal" compound needs the carboxylic function that confers the optimal acidity and the negative charge, the chiral center that allows the proper spatial disposition of the molecule, the chlorophenoxy moiety that might interact with a hydrophobic pocket (Liantonio *et al.*, 2003). However, all these compounds do not have a therapeutic interest but have allowed a better understanding of CLC-1 function. Recent studies have shown that fenofibrate *in vivo* treatment determined a reduction of gCl (Pierno *et al.*, 2006) and this reduction is caused by a direct block of the hCLC-1 channel expressed in HEK cells (Pierno *et al.*, 2007). Other effective drugs used in therapy to reduce cholesterol elevation, such as statins, have been found to induce a gCl abnormal reduction certainly implicated in the generation of drug-related side effects (Pierno *et al.*, 2006). The mechanism is currently under study, but preliminary results indicate an indirect block of the channel through the activation of PKC (Pierno *et al.*, 2007). Natural compounds such as ghrelin and GH secretagogues (i.e., hexarelin and L-163255), through a direct modulation of a specific muscular receptor, can decrease gCl by acting through a PKC-mediated modulatory pathway (Pierno *et al.*, 2003). Interestingly, niflumic acid (NFA), an NSAID, has been found to directly block CLC-1 by interacting with a binding site on channel protein. In addition, it decreases muscle gCl both directly ($IC_{50} = 42\ \mu M$) and through a PKC-mediated action due to mobilization of intracellular Ca^{2+} (Liantonio *et al.*, 2007). Also in this case, the mechanism can lead to unwanted muscular effects on chronic use of the drug.

2. CLC-K channels

The elucidation of the role of CLC-K channels in kidney salt reabsorption, obtained by using mouse models, human molecular genetics, and heterologous expression systems, has brought up a growing interest toward the identification of specific ligands that allow pharmacological interventions aimed to modulate CLC-K channel activity.

In contrast to the muscle CLC-1 channel, the pharmacological characterization of renal CLC-K channels in native systems is rather poor, mainly due to the technical difficulties in measuring the Cl^- conductance *in situ* (Nissant *et al.*, 2004; Teulon *et al.*, 2005). The limited information obtained from mouse native kidney was counterbalanced by a detailed pharmacological investigation performed on CLC-K channels expressed in heterologous systems. Indeed, after screening a variety of molecules belonging to different structural classes, it was demonstrated in the last years that CLC-K channels have two functionally different extracellular drug-binding sites: a blocking site and an activating site.

The first CLC-K pharmacological characterization has been performed on CLC-K chimeras and on rat CLC-K1 by using bis-phenoxy like compounds, derivatives of the CPP, a specific ligand of muscle CLC-1 (Liantonio *et al.*, 2002, 2004). Among several tested compounds, the CPP analogue carrying a benzyl group on the chiral center (3-phenyl-CPP) represented the minimal structure capable of stereoselectively inhibiting CLC-K1 currents with micromolar affinity ($\sim$100 μM). In particular, for CLC-Ka and CLC-K1, it was found that the block by 3-phenyl-CPP was quickly reversible and competitive with extracellular Cl^-, suggesting that the binding site for the compound is exposed to the extracellular side and is located to the ion-conducting pore. The extracellular modulation of CLC-K channels could be of high therapeutic interest, because the binding site is easily accessible, considering that the channels show specific basolateral, and possibly also apical, localization in renal epithelia.

Interestingly and unexpectedly, in view of the high homology between the two human isoforms, the apparent affinity of human CLC-Kb for 3-phenyl-CPP was found to be fivefold to sixfold lower than CLC-Ka. The use of sequence comparison and of the crystal structure of the bacterial StCLC as a guide allowed us to identify a critical residue at position 68 as the major molecular determinant for the differential behavior, as CLC-Ka has a neutral asparagines at this position whereas CLC-Kb has a charged aspartate (Picollo *et al.*, 2004).

Furthermore, the use of NFA derivatives gave the chance to unmask an activating binding site on both human CLC-K isoforms. Indeed, NFA enhances the activity of CLC-Ka and even more that of CLC-Kb. It seems that the NFA-mediated current potentiation is caused by the interaction with an activating binding site different from the blocking binding site (Liantonio *et al.*, 2006; Picollo *et al.*, 2004, 2007).

More recently, by using computational modeling, we correlated the coplanarity/noncoplanarity configuration of the aromatic portions of the molecules to the activating/blocking effect. Indeed, we showed that triflocine, the 4-aminopyridine ring analogue of NFA, shows a noncoplanar arrangement of the aromatic rings and a blocking activity similar to FFA. Conversely, GF-166, a cyclized FFA homologue, with a planar structure, exhibited an activation behavior like NFA. Thus, a coplanar conformation of the aromatic moieties favors the interaction with the activating binding site, while a noncoplanar conformation is required for a high-affinity binding with a blocking binding site. Furthermore, the same experimental approach allowed us to reveal the family of benzofuran compounds as structurally novel and general blockers of CLC-K channels. Indeed, benzofuran-like compounds, rationally designed via cyclization of 3-phenyl-CPP, resulted efficient CLC-Ka blockers with a gain of affinity of about 10-fold with respect to 3-phenyl-CPP and resulted also capable of inhibiting CLC-Kb with a K_D of 7 μM, thus representing the first blockers for this human CLC-K isoform (Liantonio *et al.*, 2008).

Taking into account the involvement of CLC-K channels in genetic and acquired diseases, these molecules could be of high therapeutic potential. Other than representing promising candidates as lead compounds for the design of new diuretic drugs (Fong, 2004), benzofuran derivatives could be of therapeutic interest also for the various form of volume-dependent/salt-sensitive hypertension (Barlassina *et al.*, 2007; Jeck *et al.*, 2004) due to gain-of-function polymorphisms within the *CLCNKA*, *CLCNKB*, or *BSND* genes. Furthermore, high-affinity CLC-K inhibitors may permit the pharmacological creation of animal models that mimic Bartter's syndrome type III or IV. Such an animal model would be of great value for elucidating the physiopathology of tubule disease in Bartter's syndrome as well as for evaluating the efficacy of possible therapies.

At the same time, it is well known that, secondary to the compromised CLC-Kb channel activity, type III Bartter's syndrome patients showed a markedly elevated prostaglandin (PGE2) activity (Reinalter *et al.*, 2002). Thus, the direct action of NFA on CLC-K channels as an opener together with its cyclooxygenase inhibition activity makes NFA a lead starting point molecule on which to work for identifying drugs that might be therapeutically useful for this renal channelopathy.

3. CLC-5 channels

Up to day any attempt to perform a pharmacological characterization failed. Indeed, several classic Cl^- channel inhibitors (DIDS, 9-AC, etc.) had no effect on heterologously expressed CLC-5. Also, CPP-like compounds, drugs active on other members of the CLC family, failed to inhibit it (Pusch *et al.*, 2000). It could be probably attributed to the fact that, as recently assumed (Picollo and Pusch, 2005; Scheel *et al.*, 2005), CLC-5 is not a Cl^- channel but rather a transporter in which the inward movement of Cl^- is stochiometrically coupled to the outward movement of H^+. Hopefully, very recently, it was suggested that benzofuran derivatives could represent a starting point structure for developing high-affinity CLC-5 ligands (Liantonio *et al.*, 2008).

4. CLC-7 channels

The CLC-7 channel has so far escaped any attempt to the biophysical and pharmacological characterization, because it does not localize in the plasma membrane in heterologous system. In contrast to this experimental limit, the therapeutic interest in identifying CLC-7-specific ligands is very high. Indeed, other than potential application of CLC-7 channel activators in the osteopetrosis treatment, specific inhibitors of CLC-7 might be useful in treating osteoporosis, a very common disorders of mostly elderly woman. Thus, attempts are ongoing to develop such inhibitors by using alternative indirect experimental strategies. The compound NS3736 and its derivatives, all belonging to the group of acidic diphenylureas, block acidification in resorption compartments and inhibit osteoclastic resorption *in vitro* (Karsdal *et al.*, 2005; Schaller *et al.*, 2004). However, it remains to verify whether these compounds act directly on CLC-7.

E. Ligand-gated ion channel receptors

The ligand-gated ion channel receptors are a group of transmembrane proteins physiologically modulated by the binding of a neurotransmitter. The GABA and glycine receptors ($GABA_AR$ and GlyR) are the major inhibitory neurotransmitter-gated receptors in the CNS. Different subunits are produced by different genes or splice variants. The receptor contains two important sites: a binding site for neurotransmitter and an ion-conducting pore. The binding of neurotransmitter induces influx of Cl^- ions within the cell which hyperpolarize the postsynaptic membrane, resulting in neurotransmitter inhibition. Dysfunction of both receptors has been implicated in channelopathies. GlyR mutations induce hyperekplexia in humans with loss of function of the channel and $GABA_AR$ mutations lead to idiopathic generalized epilepsy and Angelman's syndrome. The GlyR has a very modest pharmacological profile since no specific ligands are known. Drugs used in therapy include $GABA_AR$ agonists and 5-HT3 antagonists as tropisetron (Webb and Lynch, 2007). In contrast, the drug acting on $GABA_AR$ comprises an extensive therapeutic field, including anxiety, epilepsy, mood disorders, sleep disorders, schizophrenia, cognitive disorders, and general anesthesia (Jacob *et al.*, 2008). A mutation in α_3-subunit was found in a patient with chronic insomnia; thus, drugs acting preferentially on this subunit may be useful in the treatment of insomnia. Positive allosteric $GABA_AR$ modulators, such as phenobarbital and benzodiazepines, are widely used in the treatment of partial and generalized epilepsies. Mutations in α-, β-, and γ-subunits result in diminished synaptic inhibition. Some mutations may also impair $GABA_AR$ expression or benzodiazepine responsiveness. Other allosteric sites bind barbiturates, etomidate, *n*-octanol, ethanol, propofol, halothane, and neuroactive steroids with increase of GABAergic neurotransmission. Taurine exerts beneficial effects in the modulation of both ligand-gated receptors (Olive, 2002; Walz, 2002).

Neurotransmitter-gated cationic channels include nicotinic ACh receptor, glutamate receptor, and serotoninergic receptors. The binding of acetylcholine to its receptor induces an opening of the channel and an inward Na^+ current is responsible for generation of AP. Several nicotinic receptor subtypes show different pharmacological responses to drugs and toxins (such as nicotine, bungarotoxin, succinylcholine, tubocurarine, pancuronium). Drugs targeting nicotinic receptors are used as ganglioplegic and muscle relaxant in anesthesia. Loss-of-function mutations of nicotinic receptor cause CMS, epilepsy (Celesia, 2001; Engel, 2007) and autoimmune MG (Vernino, 2007). Recent involvement of altered AChR function has been described in Parkinson disease (PD), Alzheimer's disease (AD), and schizophrenia (Kalamida *et al.*, 2007). Drugs used in those therapies are symptomatic in the restoration of normal channel function. For instance, 3,4-diaminopyridine, which increases ACh release, and cholinesterase inhibitors have been found to be effective in solving CMS symptoms (Engel, 2007). Open-channel blockers of the AChR, quinidine, and fluoxetine are useful when the synaptic response is increased (Kalamida *et al.*, 2007). A number of nAChR agonists have shown beneficial effects in PD and AD (Kihara *et al.*, 2001; Schneider *et al.*, 1998). New drugs such as altinicline, a nicotine analog, enter the phase II for Parkinson treatment (Cosford *et al.*, 2000). The antiepileptic carbamazepine has been proposed for the autosomal-dominant nocturnal frontal lobe epilepsy (ADNFLE) characterized by nAChR α-subunit mutation (Heron *et al.*, 2007; Picard *et al.*, 1999). Lobeline, a natural nicotinic receptor agonist, and other related drugs (varenicline and epibatidine) have a potential role in smoking cessation therapy and some CNS disorders (Arneric *et al.*, 2007; Damaj *et al.*, 1997; Hays *et al.*, 2008). The serotoninergic receptor channel (5-HT3 type) is expressed in the peripheral nerve and area postrema where it plays a role in antinociception and antiemetic responses. Drugs targeting this channel (such as ondansetron and tropisetron) are in use as antiemetic. 5-HT3 receptor antagonists can also be used as analgesics (tropisetron, dolasetron) (Piper *et al.*, 2001). Glutamate receptor channels are important targets for neuroprotective and antiepileptic drugs and thereby used in the neuronal channelopathies. Drugs acting on glutamate receptor reduce excitotoxicity in the CNS being effective in acute hypoxic–ischemic brain injury and in chronic neurodegenerative diseases such as AD, PD, Hungtington's disease, and amyotrophic lateral sclerosis. Glutamate antagonists (i.e., memantine) are also effective in neuropathic pain, dementia and melanoma as well as in neuroprotection (Chen and Lipton, 2006). More recently, the use of glutamate agonists has been proposed in schizophrenia (Fell *et al.*, 2008; Imre, 2007).

IV. CONCLUSIONS

In this review, we described the mainstay therapy of ion channel diseases, especially those arising from genetic alteration of ion channel function. Although recent development of new drugs has considerably ameliorated the therapy, ion channelopathies remain a major health problem characterized by insufficient illness relief or no response to drugs at all, and by high rate of adverse events. The status of rare diseases complicates the development of large randomized clinical trials, although some tentative is ongoing. Most of the drugs were introduced in therapy based on the experience acquired quite empirically with less rare diseases with similar symptoms, and many were discovered afterward by chance to target ion channels. Intense research is being conducted to develop new drugs acting on ion channels and aimed at the understanding of the intimate drug–channel interaction. Part of this research regarding voltage-gated and neurotransmitter-gated ion channels is described in this review, whereas the molecular pharmacology of other channels (e.g., the cystic fibrosis transmembrane conductance regulator Cl^- channel, the transient receptor potential channels, etc.) can be found elsewhere. The drug–channel interactions are not straightforward as much as for "classical" receptor antagonists or enzyme inhibitors, because many drugs can affect channel gating differentially, which may be of great relevance for their usefulness in therapy. Identification of guilty mutations has contributed to the understanding of the differential role of ion channel subtypes in diseases. Except rare exceptions, selective ligands are, however, lacking and a great effort of academy and industry researchers is dedicated to the identification of such molecules, which may present an obvious advantage in increasing efficacy and limiting side effects. Such selective compounds would also be useful to definitely confirm the results obtained with genetic manipulation (knockout or knockdown experiments), the interpretation of which may be biased by residual channel expression or compensatory mechanisms. All together, these results would increase our knowledge about the contribution of each channel subtype to diseases, and greatly contribute to further development of more efficient and safer pharmacotherapy. In general, ion channel mutations can easily be studied in heterologous systems of expression. Although these systems may not mimic exactly the behavior of the channel *in vivo*, such studies have often contributed to the elucidation of the gating defect responsible for the disease, which represents the ideal target for efficient drugs. Also, *in vitro* studies have evidenced how the mutations may reduce or increase sensitivity to drugs. At this regard, the concretization of pharmacogenetics from the bench side to the clinics represents an important challenge for the next years. In the case of complex, multigenic disorders, pharmacogenomics studies would help at the understanding of the etiology and identification of new druggable targets.

Acknowledgments

This work on ion channelopathies was supported by grants from Telethon-Italy (Grant no. GGP04140), the Italian "Ministero dell'Istruzione, dell'Università e della Ricerca" (Grants no. FIRBRBNE01XMP4 and PRIN-2005059597), and the Italian Space Agency (Project OSMA). The authors declare no conflict of interest.

References

Accardi, A., and Pusch, M. (2003). Conformational changes in the pore of CLC-0. *J. Gen. Physiol.* **122,** 277–293.

Accurso, F. J. (2008). Update in cystic fibrosis 2007. *Am. J. Respir. Crit. Care Med.* **177,** 1058–1061.

Annane, D., Moore, D. H., Barnes, P. R., and Miller, R. G. (2006). Psychostimulants for hypersomnia (excessive daytime sleepiness) in myotonic dystrophy. *Cochrane Database Syst. Rev.* **3,** **CD003218.**

Argov, Z., McKee, D., Agus, S., Brawer, S., Shlomowitz, N., Yoseph, O. B., Soreq, H., and Sussman, J. D. (2007). Treatment of human myasthenia gravis with oral antisense suppression of acetylcholinesterase. *Neurology* **69,** 699–700.

Arneric, S. P., Holladay, M., and Williams, M. (2007). Neuronal nicotinic receptors: A perspective on two decades of drug discovery research. *Biochem. Pharmacol.* **74,** 1092–1101.

Arnestad, M., Crotti, L., Rognum, T. O., Insolia, R., Pedrazzini, M., Ferrandi, C., Vege, A., Wang, D. W., Rhodes, T. E., George, A. L., Jr., and Schwartz, P. J. (2007). Prevalence of Long-QT syndrome gene variants in sudden infant death syndrome. *Circulation* **115,** 361–367.

Aromataris, E. C., Astill, D. S., Rychkov, G. Y., Bryant, S. H., Bretag, A. H., and Roberts, M. L. (1999). Modulation of the gating of ClC-1 by S-(−) 2-(4-chlorophenoxy)propionic acid. *Br. J. Pharmacol.* **126,** 1375–1382.

Ashcroft, F. M. (2006). From molecule to malady. *Nature* **440,** 440–447.

Askmyr, M. K., Fasth, A., and Richter, J. (2008). Towards a better understanding and new therapeutics of osteopetrosis. *Br. J. Haematol.* **140,** 597–609.

Attal, N., Cruccu, G., Haanpaa, M., Hansson, P., Jensen, T. S., Numikko, T., Sampaio, C., Sindrup, S., and Wilfen, P. (2006). EFNS guidelines on pharmacological treatment of neuropathic pain. *Eur. J. Neurol.* **13,** 1153–1169.

Ayata, C., Hongwei, J., Kudo, C., Dalkara, T., and Moskowitz, M. A. (2006). Suppression of cortical spreading depression in migraine prophylaxis. *Ann. Neurol.* **59,** 652–661.

Bakker, M. J., van Dijk, J. G., van den Maagdenberg, A. M. J. M., and Tijssen, M. A. (2006). Startle syndromes. *Lancet Neurol.* **5,** 513–524.

Barlassina, C., Dal Fiume, C., Lanzani, C., Manunta, P., Guffanti, G., Ruello, A., Bianchi, G., Del Vecchio, L., Macciardi, F., and Cusi, D. (2007). Common genetic variants and haplotypes in renal CLCNKA gene are associated to salt-sensitive hypertension. *Hum. Mol. Genet.* **16,** 1630–1638.

Belhassen, B., Glick, A., and Viskin, S. (2004). Efficacy of quinidine in high-risk patients with Brugada syndrome. *Circulation* **110,** 1731–1737.

Bellinger, A. M., Mongillo, M., and Marks, A. R. (2008). Stressed out: The skeletal muscle ryanodine receptor as a target of stress. *J. Clin. Invest.* **118,** 445–453.

Benhorin, J., Taub, R., Goldmit, M., Kerem, B., Kass, R. S., Windman, I., and Medina, A. (2000). Effects of flecainide in patients with new SCN5A mutation: Mutation-specific therapy for long-QT syndrome? *Circulation* **101,** 1698–1706.

Benos, D. J. (1982). Amiloride: A molecular probe of sodium transport in tissues and cells. *Am. J. Physiol.* **242,** C131–C145.

Bezzina, C. R., and Tan, H. L. (2002). Pharmacological rescue of mutant ion channels. *Cardiovasc. Res.* **55,** 229–232.

Bien, C. G., Granata, T., Antozzi, C., Cross, J. H., Dulac, O., Kurthen, M., Lassmann, H., Mantegazza, R., Villemure, J. G., Spreafico, R., and Elger, C. E. (2005). Pathogenesis, diagnosis and treatment of Rasmussen encephalitis: A European consensus statement. *Brain* **128,** 454–471.

Birkenhäger, R., Otto, E., Schurmann, M. J., Vollmer, M., Ruf, E.-M., Maier-Lutz, I., Beekmann, F., Fekete, A., Omran, H., Feldmann, D., Milford, D. V., Jeck, N., *et al.* (2001). Mutation of BSND causes Bartter syndrome with sensorineural deafness and kidney failure. *Nat. Genet.* **29,** 310–314.

Boneva, N., Frenkian-Cuvelier, M., Bidault, J., Brenner, T., and Berrih-Aknin, S. (2006). Major pathogenic effects of anti-MuSK antibodies in myasthenia gravis. *J. Neuroimmunol.* **177,** 119–131.

Bongioanni, P., Ricciardi, R., and Romano, M. R. (1999). T-lymphocyte interferon-gamma receptor binding in patients with myasthenia gravis. *Arch. Neurol.* **56,** 933–938.

Botta, A., Vallo, L., Rinaldi, F., Bonifazi, E., Amati, F., Biancolella, M., Gambardella, S., Mancinelli, E., Angelini, C., Meola, G., and Novelli, G. (2007). Gene expression analysis in myotonic dystrophy: Indications for a common molecular pathogenic pathway in DM1 and DM2. *Gene Empr.* **13,** 339–351.

Bryant, S. H., and Conte Camerino, D. (1991). Chloride channel regulation in the skeletal muscle of normal and myotonic goats. *Pflügers Arch.* **417,** 605–610.

Bryant, S. H., and Morales-Aguilera, A. (1971). Chloride conductance in normal and myotonic muscle fibres and the action of monocarboxylic aromatic acids. *J. Physiol.* **219,** 367–383.

Buckley, C., and Vincent, A. (2005). Autoimmune channelopathies. *Nat. Clin. Pract. Neurol.* **1,** 22–33.

Cebotaru, V., Kaul, S., Devuyst, O., Cai, H., Racusen, L., Guggino, W. B., and Guggino, S. E. (2005). High citrate diet delays progression of renal insufficiency in the ClC-5 knockout mouse model of Dent's disease. *Kidney Int.* **68,** 642–652.

Celesia, G. G. (2001). Disorders of membrane channels or channelopathies. *Clin. Neurophysiol.* **112,** 12–18.

Chaitman, B. R. (2006). Ranolazine for the treatment of chronic angina and potential se in other cardiovascular conditions. *Circulation* **113,** 2462–2472.

Chalhoub, N., Benachenhou, N., Rajapurohitam, V., Pata, M., Ferron, M., Frattini, A., Villa, A., and Vacher, J. (2003). Grey-lethal mutation induces severe malignant autosomal recessive osteopetrosis in mouse and human. *Nat. Med.* **9,** 399–406.

Chang, S. S., Grunder, S., Hanukoglu, A., Rösler, A., Mathew, P. M., Hanukoglu, I., Schild, L., Lu, Y., Shimkets, R. A., Nelson-Williams, C., Rossier, B. C, and Lifton, R. P (1996). Mutations in subunits of the epithelial sodium channel cause salt wasting with hyperkalaemic acidosis, pseudohypoaldosteronism type 1. *Nat Genet.* **12,** 248–253.

Chen, H. S., and Lipton, S. A. (2006). The chemical biology of clinically tolerated NMDA receptor antagonists. *J. Neurochem.* **97,** 1611–1626.

Chen, M. F., Niggeweg, R., Iaizzo, P. A., Lehmann-Horn, F., and Jockusch, H. (1997). Chloride conductance in mouse muscle is subject to post-transcriptional compensation of the functional Cl^- channel 1 gene dosage. *J. Physiol.* **504,** 75–81.

Chen, Y., Lu, J., Pan, H., Zhang, Y., Wu, H., Xu, K., Liu, X., Jiang, Y., Bao, X., Yao, Z., Ding, K., Lo, W. H., *et al.* (2003). Association between genetic variation of CACNA1H and childhood absence epilepsy. *Ann. Neurol.* **54,** 239–243.

Christie, S., Gobel, H., Mateos, V., Allen, C., and Vrijens, F. (2003). Crossover comparison of efficacy and preference for rizatriptan 10 mg versus ergotamine/caffeine in migraine. *Eur. Neurol.* **49**(1), 20–29.

Clare, J. J., Tate, S. N., Nobbs, M., and Romanos, M. A. (2000). Voltage-gated sodium channels as therapeutic targets. *Drug Discov. Today* **5,** 506–520.

Cohen-Kaminsky, S., and Jambou, F. (2005). Prospects for a T-cell receptor vaccination against myasthenia gravis. *Expert Rev. Vaccines* **4,** 473–492.

Cole, R. N., Reddel, S. W., Gervásio, O. L., and Phillips, W. D. (2008). Anti-MuSK patient antibodies disrupt the mouse neuromuscular junction. *Ann. Neurol.* **63,** 782–789.

Colman, J., Rothney, A., Wright, S. C., Zilkalns, B., and Rowe, R. H. (2004). Use of narcotic analgesics in the emergency department treatment of migraine headache. *Neurology* **62,** 1695–1700.

Conte Camerino, D., Tortorella, V., Ferranini, E., and Bryant, S. H. (1984). The toxic effects of clofibrate and its metabolite on mammalian skeletal muscle: An electrophysiological study. *Arch. Toxicol. Suppl.* **7,** 482–484.

Conte Camerino, D., De Luca, A., Mambrini, M., Ferrannini, E., Franconi, F., Giotti, A., and Bryant, S. H. (1989). The effects of taurine on pharmacologically induced myotonia. *Muscle Nerve* **12,** 898–904.

Conte Camerino, D., Tricarico, D., Pierno, S., Desaphy, J. F., Liantonio, A., Pusch, M., Burdi, R., Camerino, C., Fraysse, B., and De Luca, A. (2004). Taurine and skeletal muscle disorders. *Neurochem. Res.* **29,** 135–142.

Conte Camerino, D., Tricarico, D., and Desaphy, J. F. (2007). Ion channel pharmacology. *Neurotherapeutics* **4,** 184–198.

Conte-Camerino, D., Mambrini, M., De Luca, A., Tricarico, D., Bryant, S. H., Tortorella, V., and Bettoni, G. (1988). Enantiomers of clofibric acid analogs have opposite actions on rat skeletal muscle chloride channels. *Pflügers Arch.* **413,** 105–107.

Cosford, N. D., Bleicher, L., Vernier, J. M., Chavez-Noriega, L., Rao, T. S., Siegel, R. S., Suto, C., Washburn, M., Lloyd, G. K., and McDonald, I. A. (2000). Recombinant human receptors and functional assays in the discovery of altinicline (SIB-1508Y), a novel acetylcholine-gated ion channel (nAChR) agonist. *Pharm. Acta Helv.* **74,** 125–130.

Cragg, M. S., Walshe, C. A., Ivanov, A. O., and Glennie, M. J. (2005). The biology of CD20 and its potential as a target for mAb therapy. *Curr. Dir. Autoimmun.* **8,** 140–174.

Dalakas, M. C. (2008). Advances in the pathogenesis and treatment of patients with stiff person syndrome. *Curr. Neurol. Neurosci. Rep.* **8,** 48–55.

Dalmau, J., Graus, F., Villarejo, A., Posner, J. B., Blumenthal, D., Thiessen, B., Saiz, A., Meneses, P., and Rosenfeld, M. R. (2004). Clinical analysis of anti-Ma2-associated encephalitis. *Brain* **127,** 1831–1844.

Damaj, M. I., Patrick, G. S., Creasy, K. R., and Martin, B. R. (1997). Pharmacology of lobeline, a nicotinic receptor ligand. *J. Pharmacol. Exp. Ther.* **282,** 410–419.

De Giorgio, C. M., Rabinowicz, A. L., and Olivas, R. D. (1991). Carbamazepine-induced antinuclear antibodies and systemic lupus erythematosus-like syndrome. *Epilepsia* **32,** 128–129.

De Luca, A., Tricarico, D., Wagner, R., Bryant, S. H., Tortorella, V., and Conte Camerino, D. (1992). Opposite effects of enantiomers of clofibric acid derivative on rat skeletal muscle chloride conductance: Antagonism studies and theoretical modeling of two different receptor site interactions. *J. Pharmacol. Exp. Ther.* **260,** 364–368.

De Luca, A., Tricarico, D., Pierno, S., and Conte Camerino, D. (1994). Aging and chloride channel regulation in rat fast-twitch muscle fibres. *Pflügers Arch.* **427,** 80–85.

De Luca, A., Natuzzi, F., Desaphy, J. F., Loni, G., Lentini, G., Franchini, C., Tortorella, V., and Conte Camerino, D. (2000). Molecular determinants of mexiletine structure for potent and use-dependent block of skeletal muscle sodium channels. *Mol. Pharmacol.* **57,** 268–277.

De Luca, A., Talon, S., De Bellis, M., Desaphy, J. F., Lentini, G., Corbo, F., Scilimiati, A., Franchini, C., Tortorella, V., and Conte Camerino, D. (2003). Optimal requirements for high affinity and use dependent block of skeletal muscle sodium channel by *N*-benzyl analogs of tocainide-like compounds. *Mol. Pharmacol.* **64,** 932–945.

Del Fattore, A., Cappariello, A., and Teti, A. (2008). Genetics, pathogenesis and complications of osteopetrosis. *Bone* **42,** 19–29.

Desaphy, J. F., Conte Camerino, D., Franchini, C., Lentini, G., Tortorella, V., and De Luca, A. (1999). Increased hindrance on the chiral carbon atom of mexiletine enhances the block of rat skeletal muscle Na channels in a model of myotonia induced by ATX. *Br. J. Pharmacol.* **128,** 1165–1174.

Desaphy, J. F., De Luca, A., Tortorella, P., De Vito, D., George, A. L., Jr., and Conte Camerino, D. (2001). Gating of myotonic Na channel mutants defines the response to mexiletine and a potent derivative. *Neurology* **57,** 1849–1857.

Desaphy, J. F., Pierno, S., De Luca, A., Didonna, M. P., and Conte Camerino, D. (2003). Different ability of clenbuterol and salbutamolo to block sodium channels predicts their therapeutic use in muscle excitability disorders. *Mol. Pharmacol.* **63,** 659–670.

Desaphy, J. F., De Luca, A., Didonna, M. P., George, A. L., Jr., and Conte Camerino, D. (2004). Different flecainide sensitivity of hNav1.4 channels and myotonic mutants explained by state-dependent block. *J. Physiol.* **554,** 321–334.

Desaphy, J. F., Rolland, J. F., Valente, E. M., LoMonaco, M., and Conte Camerino, D. (2007). Functional alteration of ClC-1 channel mutants associated with transient weakness in myotonia congenita. *Biophys. J.* **92,** 273a (abstract).

Deterding, R. R., Lavange, L. M., Engels, J. M., Mathews, D. W., Coquillette, S. J., Brody, A. S., Millard, S. P., and Ramsey, B. W. (2007). Phase 2 randomized safety and efficacy trial of nebulized denufosol tetrasodium in cystic fibrosis. *Am. J. Respir. Crit. Care Med.* **176,** 362–369.

Diener, H. C., Jansen, J. P., Reches, A., Pascual, J., Pitei, D., and Steiner, T. J. (2002). Efficacy, tolerability, and safety of oral eletriptan and ergotamine plus caffeine (Cafergot) in the acute treatment of migraine: A multicenter, randomized, double-blind, placebo-controlled comparison. *Eur. Neurol.* **47,** 99–107.

Dooley, D. J., Taylor, C. P., Donevan, S., and Feltner, D. (2007). Ca^{2+} channel $\alpha 2\delta$ ligands: Novel modulators of neurotransmission. *Trends Pharmacol. Sci.* **28,** 75–82.

Dörner, T., and Goldenberg, D. M. (2007). Targeting CD22 as a strategy for treating systemic autoimmune diseases. *Ther. Clin. Risk Manag.* **3,** 953–959.

Drachman, D. B. (1994). Myasthenia gravis. *N. Engl. J. Med.* **330,** 1797–1810.

Drachman, D. B., Wu, J. M., Miagkov, A., Williams, M. A., Adams, R. N., and Wu, B. (2003). Specific immunotherapy of experimental myasthenia by genetically engineered APCs: The "guided missile" strategy. *Ann. N.Y. Acad. Sci.* **998,** 520–532.

Drenth, J. P. H., and Waxman, S. G. (2007). Mutations in sodium-channel gene SCN9A cause a spectrum of human genetic pain disorders. *J. Clin. Invest.* **117,** 3603–3609.

Dworkin, R. H., O'Connor, A. B., Backonja, M., Farrar, J. T., Finnerup, N. B., Jensen, T. S., Kalso, E. A., Loeser, J. T., Miaskowski, C., Nurmikko, T. J., *et al.* (2007). Pharmacologic management of neuropathic pain: Evidence-based recommendations. *Pain* **132,** 237–250.

Eguchi, H., Tsujino, A., Kaibara, M., Hayashi, H., Shirabe, S., Taniyama, K., and Eguchi, K. (2006). Acetazolamide acts directly on the human skeletal muscle chloride channel. *Muscle Nerve* **34,** 292–297.

Engel, A. G. (2007). The therapy of congenital myasthenic syndromes. *Neurotherapeutics* **4,** 252–257.

Farrugia, M. E., Robson, M. D., Clover, L., Anslow, P., Newsom-Davis, J., Kennett, R., Hilton-Jones, D., Matthews, P. M., and Vincent, A. (2006). MRI and clinical studies of facial and bulbar muscle involvement in MuSK antibody-associated myasthenia gravis. *Brain* **129,** 1481–1492.

Fell, M. J., Johnson, B. G., Svensson, K. A., and Schoepp, D. D. (2008). Evidence for the role of mGlu2 not mGlu3 receptors in the pre-clinical antipsychotic pharmacology of the mGlu2/3 receptor agonist LY404039. *J. Pharmacol. Exp. Ther.* **326,** 209–217.

Fertleman, C. R., Baker, M. D., Parker, K. A., Moffatt, S., Elmslie, F. V., Abrahamsen, B., Ostman, J., Klugbauer, N., Wood, J. N., Gardiner, R. M., and Rees, M. (2006). SCN9A mutations in paroxysmal extreme pain disorder: Allelic variants underlie distinct channel defects and phenotypes. *Neuron* **52,** 757–774.

Fong, P. (2004). CLC-K channels: If the drug fits, use it. *EMBO Rep.* **5,** 565–566.

Gazulla, J., and Tintoré, M. (2007). The P/Q-type voltage-dependent calcium channel: A therapeutic target in spinocerebellar ataxia type 6. *Acta Neurol. Scand.* **115,** 356–363.

Grasemann, H., Stehling, F., Brunar, H., Widmann, R., Laliberte, T. W., Molina, L., Döring, G., and Ratjen, F. (2007). Inhalation of Moli1901 in patients with cystic fibrosis. *Chest* **131,** 1461–1466.

Gründer, S., Firsov, D., Chang, S. S., Jaeger, N. F., Gautschi, I., Schild, L., Lifton, R. P., and Rossier, B. C. (1997). A mutation causing pseudohypoaldosteronism type 1 identifies a conserved glycine that is involved in the gating of the epithelial sodium channel. *EMBO J.* **16,** 899–907.

Gultekin, S. H., Rosenfeld, M. R., Voltz, R., Eichen, J., Posner, J. B., and Dalmau, J. (2000). Paraneoplastic limbic encephalitis: Neurological symptoms, immunological findings and tumour association in 50 patients. *Brain* **123,** 1481–1494.

Hansson, J. H., Schild, L., Lu, Y., Wilson, T. A., Gautschi, I., Shimkets, R., Nelson-Williams, C., Rossier, B. C., and Lifton, R. P. (1995). A *de novo* missense mutation of the beta subunit of the epithelial sodium channel causes hypertension and Liddle syndrome, identifying a proline-rich segment critical for regulation of channel activity. *Proc. Natl Acad. Sci. USA* **92,** 11495–11499.

Harrower, T., Foltynie, T., Kartsounis, L., De Silva, R. N., and Hodges, J. R. (2006). A case of voltage-gated potassium channel antibody-related limbic encephalitis. *Nat. Clin. Pract. Neurol.* **2,** 339–343.

Hart, I. K., Maddison, P., Newsom-Davis, J., Vincent, A., and Mills, K. R. (2002). Phenotypic variants of autoimmune peripheral nerve hyperexcitability. *Brain* **125,** 1887–1895.

Harteneck, C. (2005). Function and pharmacology of TRPM cation channels. *Naunyn-Schmied. Arch. Pharmacol.* **371,** 307–314.

Hays, J. T., Ebbert, J. O., and Sood, A. (2008). Efficacy and safety of varenicline for smoking cessation. *Am. J. Med.* **121**(Suppl. 1), S32–S42.

Heatwole, C., and Moxley, R. T., III. (2007). The nondystrophic myotonias. *Neurotherapeutics* **4,** 238–251.

Hebert, S. C. (2003). Bartter syndrome. *Curr. Opin. Nephrol. Hypertens.* **12,** 527–532.

Heron, S. E., Scheffer, I. E., Berkovic, S. F., Dibbens, L. M., and Mulley, J. C. (2007). Channelopathies in idiopathic epilepsy. *Neurotherapeutics* **4,** 295–304.

Hille, B. (2001). "Ion Channels of Excitable Membranes," 3rd edn. Sinauer Associates, Sunderland.

Hirsh, A. J., Molino, B. F., Zhang, J., Astakhova, N., Geiss, W. B., Sargent, B. J., Swenson, B. D., Usyatinsky, A., Wyle, M. J., Boucher, R. C., *et al.* (2006). Design, synthesis, and structure–activity relationships of novel 2-substituted pyrazinoylguanidine epithelial sodium channel blockers: Drugs for cystic fibrosis and chronic bronchitis. *J. Med. Chem.* **49,** 4098–4115.

Hoenderop, J. G., and Bindels, R. J. (2005). Epithelial Ca^{2+} and Mg^{2+} channels in health and disease. *J. Am. Soc. Nephrol.* **16,** 15–26.

Holland, K. D., Keamey, J. A., Glauser, T. A., Buck, G., Keddache, M., Blankston, J. R., Glaaser, I. W., Kass, R. S., and Meisler, M. H. (2008). Mutation of sodium channel SCN3A in a patient with cryptogenic pediatric partial paralysis. *Neurosci. Lett.* **433,** 65–70.

Imre, G. (2007). The preclinical properties of a novel group II metabotropic glutamate receptor agonist LY379268. *CNS Drug Rev.* **13,** 444–464.

Jacob, T. C., Moss, S. J., and Jurd, R. (2008). GABA(A) receptor trafficking and its role in the dynamic modulation of neuronal inhibition. *Nat. Rev. Neurosci.* **9,** 331–343.

Jahangir, A., and Terzic, A. (2005). KATP channel therapeutics at the bedside. *J. Mol. Cell. Cardiol.* **39,** 99–112.

Jarius, S., Paul, F., Franciotta, D., Waters, P., Zipp, F., Hohlfeld, R., Vincent, A., and Wildemann, B. (2008). Mechanisms of disease: Aquaporin-4 antibodies in neuromyelitis optica. *Nat. Clin. Pract. Neurol.* **4,** 202–214.

Jarvis, M. F., Honore, P., Shieh, C. C., Chapman, M., Joshi, S., Zhang, X. F., Kort, M., Carroll, W., Marron, B., Atkinson, R., Thomas, J., Liu, D., *et al.* (2007). A-803467, a potent and selective Nav1.8 sodium channel blocker, attenuates neuropathic and inflammatory pain in the rat. *Proc. Natl. Acad. Sci. USA* **104,** 8520–8525.

Jeck, N., Waldegger, S., Lampert, A., Boehmer, C., Waldegger, P., Lang, P. A., Wissinger, B., Friedrich, B., Risler, T., Moehle, R., Lang, U. E., Zill, P., *et al.* (2004). Activating mutation of the renal epithelial chloride channel ClC-Kb predisposing to hypertension. *Hypertension* **43,** 1175–1181.

Jentsch, T. J. (2008). CLC chloride channels and transporters: From genes to protein structure, pathology and physiology. *Crit. Rev. Biochem. Mol. Biol.* **43,** 3–36.

Jurkat-Rott, K., and Lehmann-Horn, F. (2005). Muscle channelopathies and critical points in functional and genetic studies. *J. Clin. Invest.* **115,** 2000–2009.

Jurkat-Rott, K., and Lehmann-Horn, F. (2007). Do hyperpolarization induced proton currents contribute to the pathogenesis of hypokalemic periodic paralysis, a voltage sensor channelopathy? *J. Gen. Physiol.* **130,** 1–5.

Kalamida, D., Poulas, K., Avramopoulou, V., Fostieri, E., Lagoumintzis, K., Sideri, A., Zouridakis, M., and Tzartos, S. J. (2007). Muscle and neuronal nicotinic acetylcholine receptors. Structure, function and pathogenicity. *FEBS J.* **274,** 3799–3845.

Karsdal, M. A., Henriksen, K., Sørensen, M. G., Gram, J., Schaller, S., Dziegiel, M. H., Heegaard, A. M., Christophersen, P., Martin, T. J., Christiansen, C., *et al.* (2005). Acidification of the osteoclastic resorption compartment provides insight into the coupling of bone formation to bone resorption. *Am. J. Pathol.* **166,** 467–476.

Kasper, D., Planells-Cases, R., Fuhrmann, J. C., Scheel, O., Zeitz, O., Ruether, K., Schmitt, A., Poët, M., Steinfeld, R., Schweizer, M., Kornak, U., and Jentsch, T. J. (2005). Loss of the chloride channel ClC-7 leads to lysosomal storage disease and neurodegeneration. *EMBO J.* **24,** 1079–1091.

Kazkaz, H., and Isenberg, D. (2004). Anti B cell therapy (rituximab) in the treatment of autoimmune diseases. *Curr. Opin. Pharmacol.* **4,** 398–402.

Kihara, T., Shimohama, S., Sawada, H., Honda, K., Nakamizo, T., Shibasaki, H., Kume, T., and Akaike, A. (2001). Alpha 7 nicotinic receptor transduces signals to phosphatidylinositol 3-kinase to block A beta-amyloid-induced neurotoxicity. *J. Biol. Chem.* **276,** 13541–13546.

Kleta, R., and Bockenhauer, D. (2006). Bartter syndromes and other salt-losing tubulopathies. *Nephron Physiol.* **104,** 73–80.

Konrad, M., Vollmer, M., Lemmink, H. H., van den Heuvel, L. P., Jeck, N., Vargas-Poussou, R., Lakings, A., Ruf, R., Deschênes, G., Antignac, C., Guay-Woodford, L., Knoers, N. V., *et al.* (2000). Mutations in the chloride channel gene CLCNKB as a cause of classic Bartter syndrome. *J. Am. Soc. Nephrol.* **11,** 1449–1459.

Kordasiewicz, H. B., and Gomez, C. M. (2007). Molecular pathogenesis of spinocerebellar ataxia type 6. *Neurotherapeutics* **4,** 285–294.

Kornak, U., Kasper, D., Bösl, M. R., Kaiser, E., Schweizer, M., Schulz, A., Friedrich, W., Delling, G., and Jentsch, T. J. (2001). Loss of the ClC-7 chloride channel leads to osteopetrosis in mice and man. *Cell* **104,** 205–215.

Kort, M. E., Drizin, I., Gregg, R. J., Scanio, M. J., Shi, L., Gross, M. F., Atkinson, R. N., Johnson, M. S., Pacofsky, G. J., Thomas, J. B., Carroll, W. A., Krambis, M. J., *et al.* (2008). Discovery and biological evaluation of 5-aryl-2-furfuramides, potent and selective blockers of the Nav1.8 sodium channel with efficacy in models of neuropathic and inflammatory pain. *J. Med. Chem.* **51,** 407–416.

Krause, T., Gerbershagen, M. U., Fiege, M., Weißhorn, R., and Wappler, F. (2004). Dantrolene—A review of its pharmacology, therapeutic use and new developments. *Anaesthesia* **59,** 364–373.

Kuo, C. C., Huang, R. C., and Lou, B. S. (2000). Inhibition of Na^+ current by diphenhydramine and other diphenyl compounds: Molecular determinants of selective binding to the inactivated channels. *Mol. Pharmacol.* **57,** 135–143.

Lacy, B. E., and Levy, L. C. (2007). Lubiprostone: A chloride channel activator. *J. Clin. Gastroenterol.* **41,** 345–351.

Lange, P. F., Wartosch, L., Jentsch, T. J., and Fuhrmann, J. C. (2006). ClC-7 requires Ostm1 as a beta-subunit to support bone resorption and lysosomal function. *Nature* **440,** 220–223.

Legroux-Crespel, E., Sassolas, B., Guillet, G., Kupfer, I., Dupre, D., and Misery, L. (2003). Treatment of familial erythermalgia with the association of lidocaine and mexiletine. *Ann. Dermatol. Venereol.* **130,** 429–433.

Lehnart, S. E., Ackerman, M. J., Benson, D. W., Jr., Brugada, R., Clancy, C. E., Donahue, J. K., George, A. L., Jr., Grant, A. O., Groft, S. C., January, C. T., Lathrop, D. A., Lederer, W. J., *et al.* (2007). Inherited arrhythmias: A national Heart, lung, and blood institute and office of rare diseases workshop consensus report about the diagnosis, phenotyping, molecular mechanisms, and therapeutic approaches for primary cardiomyopathies of gene mutations affecting ion channel function. *Circulation* **116,** 2325–2345.

Lenkey, N., Karoly, R., Kiss, J. P., Szasz, B. K., Vizi, E. S., and Mike, A. (2006). The mechanism of activity-dependent sodium channel inhibition by the antidepressants fluoxetine and desipramine. *Mol. Pharmacol.* **70,** 2052–2063.

Lennon, V. A., Kryzer, T. J., Pittock, S. J., Verkman, A. S., and Hinson, S. R. (2005). IgG marker of optic-spinal multiple sclerosis binds to the aquaporin-4 water channel. *J. Exp. Med.* **202,** 473–477.

Liantonio, A., Accardi, A., Carbonara, G., Fracchiolla, G., Loiodice, F., Tortorella, P., Traverso, S., Guida, P., Pierno, S., De Luca, A., Conte Camerino, D., and Pusch, M. (2002). Molecular requisites for drug binding to muscle CLC-1 and renal CLC-K channel revealed by the use of phenoxy-alkyl derivatives of 2-(p-chlorophenoxy)propionic acid. *Mol. Pharmacol.* **62,** 265–271.

Liantonio, A., De Luca, A., Pierno, S., Didonna, M. P., Loiodice, F., Fracchiolla, G., Tortorella, P., Laghezza, A., Bonerba, E., Traverso, S., Elia, L., Picollo, A., *et al.* (2003). Structural requisites of 2-(p-chlorophenoxy)propionic acid analogues for activity on native rat skeletal muscle chloride conductance and on heterologously expressed CLC-1. *Br. J. Pharmacol.* **139,** 1255–1264.

Liantonio, A., Pusch, M., Picollo, A., Guida, P., De Luca, A., Pierno, S., Fracchiolla, G., Loiodice, F., Tortorella, P., and Conte Camerino, D. (2004). Investigations of pharmacologic properties of the renal CLC-K1 chloride channel co-expressed with barttin by the use of 2-(*p*-chlorophenoxy) propionic acid derivatives and other structurally unrelated chloride channels blockers. *J. Am. Soc. Nephrol.* **15,** 13–20.

Liantonio, A., Picollo, A., Babini, E., Carbonara, G., Fracchiolla, G., Loiodice, F., Tortorella, V., Pusch, M., and Conte Camerino, D. (2006). Activation and inhibition of kidney CLC-K chloride channels by fenamates. *Mol. Pharmacol.* **69,** 165–173.

Liantonio, A., Giannuzzi, V., Picollo, A., Babini, E., Pusch, M., and Conte Camerino, D. (2007). Niflumic acid inhibits chloride conductance of rat skeletal muscle by directly inhibiting the CLC-1 channel and by increasing intracellular calcium. *Br. J. Pharmacol.* **150,** 235–247.

Liantonio, A., Picollo, A., Carbonara, G., Fracchiolla, G., Tortorella, P., Loiodice, F., Laghezza, A., Babini, E., Zifarelli, G., Pusch, M., and Conte Camerino, D. (2008). Molecular switch for CLC-K CL-channel block/activation: optimal pharmacophoric requirements towards high-affinity ligands. *Proc. Natl. Acad. Sci. USA* **105,** 1369–1373.

Liguori, R., Vincent, A., Clover, L., Avoni, P., Plazzi, G., Cortelli, P., Baruzzi, A., Carey, T., Gambetti, P., Lugaresi, E., and Montagna, P. (2001). Morvan's syndrome: Peripheral and central nervous system and cardiac involvement with antibodies to voltage-gated potassium channels. *Brain* **124,** 2417–2426.

Links, T. P., Smit, A. J., and Reitsma, W. D. (1995). Potassium channel modulation: Effect of pinacidil on insulin release in healthy volunteers. *J. Clin. Pharmacol.* **35,** 291–294.

Lipkind, G. M., and Fozzard, H. A. (2005). Molecular modeling of local anesthetic drug binding by voltage-gated sodium channels. *Mol. Pharmacol.* **68,** 1611–1622.

Lloyd, S. E., Pearce, S. H., Fisher, S. E., Steinmeyer, K., Schwappach, B., Scheinman, S. J., Harding, B., Bolino, A., Devoto, M., Goodyer, P., Rigden, S. P., Wrong, O., *et al.* (1996). A common molecular basis for three inherited kidney stone diseases. *Nature* **379,** 445–449.

Lu, M., Echeverri, F., Kalabat, D., Laita, B., Dahan, D. S., Smith, R. D., Xu, H., Staszewski, L., Yamamoto, J., Ling, J., Hwang, N., Kimmich, R., *et al.* (2008). Small molecule activator of the human epithelial sodium channel. *J. Biol. Chem.* **283,** 11981–11994.

Lucas, P. T., Meadows, L. S., Nicholls, J., and Ragsdale, D. S. (2005). An epilepsy mutation in the $\beta 1$ subunit of the voltage-gated sodium channel results in reduced channel sensitivity to phenytoin. *Epilepsy Res.* **64,** 77–84.

Ludwig, M., Utsch, B., Balluch, B., Fründ, S., Kuwertz-Bröking, E., and Bökenkamp, A. (2006). Hypercalciuria in patients with CLCN5 mutations. *Pediatr. Nephrol.* **21,** 1241–1250.

Lykke Thomsen, L., Kirchmann Eriksen, M., Faerch Romer, S., Andersen, I., Ostergaard, E., Keiding, N., Olesen, J., and Russell, M. B. (2002). An epidemiological survey of hemiplegic migraine. *Cephalalgia* **22,** 361–375.

Makielski, J. C., and Valdivia, C. R. (2006). Ranolazine and late cardiac sodium current—A therapeutic target for angina, arrhythmia and more? *Br. J. Pharmacol.* **148,** 4–6.

Makita, N., Behr, E., Shimizu, W., Horie, M., Sunami, A., Crotti, L., Schulze-Bahr, E., Fukuhara, S., Mochizuki, N., Makiyama, T., Itoh, H., Christiansen, M., *et al.* (2008). The E1784K mutation in SCN5A is associated with mixed clinical phenotype of type 3 long QT syndrome. *J. Clin. Invest.* **118,** 2219–2229.

Matsuda, S., and Koyasu, S. (2000). Mechanisms of action of cyclosporine. *Immunopharmacology* **47,** 119–125.

Miceli, F., Soldovieri, M. V., Martire, M., and Tagliatatela, M. (2008). Molecular pharmacology and therapeutic potential of neuronal Kv7-modulating drugs. *Curr. Opin. Pharmacol.* **8,** 65–74.

Mukerji, N., Damodaran, T. V., and Winn, M. P. (2007). TRPC6 and FSGS: The latest TRP channelopathy. *Biochim. Biophys. Acta* **1772,** 859–868.

Naderi, A. S., and Reilly, R. F., Jr. (2008). Hereditary etiologies of hypomagnesemia. *Nat. Clin. Pract. Nephrol.* **4,** 80–89.

Naesens, M., Steels, P., Verberckmoes, R., Vanrenterghem, Y., and Kuypers, D. (2004). Bartter's and Gitelman's syndromes: From gene to clinic. *Nephron Physiol.* **96,** 65–78.

Nichols, C. G. (2006). KATP channels as molecular sensors of cellular metabolism. *Nature* **440,** 470–476.

Nissant, A., Lourdel, S., Baillet, S., Paulais, M., Marvao, P., Teulon, J., and Imbert-Teboul, M. (2004). Heterogeneous distribution of chloride channels along the distal convoluted tubule probed by single-cell RT-PCR and patch clamp. *Am. J. Physiol. Renal. Physiol.* **287,** F1233–F1243.

Olive, M. F. (2002). Interactions between taurine and ethanol in the central nervous system. *Amino Acids* **23,** 345–357.

O'Shaughnessy, K. M., and Karet, F. E. (2004). Salt handling and hypertension. *J. Clin. Invest.* **113,** 1075–1081.

O'Shea, S. M., Becker, L., Weiher, H., Betz, H., and Laube, B. (2004). Propofol restores the function of hyperekplexic mutant glycine receptors in Xenopus oocytes and mice. *J. Neurosci.* **24,** 2322–2327.

Pascuzzi, R. M., Coslett, H. B., and Johns, T. R. (1984). Long-term corticosteroid treatment of myasthenia gravis: Report of 116 patients. *Ann. Neurol.* **15,** 291–298.

Pearson, E. R., Flechtner, I., Njølstad, P. R., Malecki, M. T., Flanagan, S. E., Larkin, B., Ashcroft, F. M., Klimes, I., Codner, E., Iotova, V., Slingerland, A. S., Shield, J., *et al.* (2006). Switching from insulin to oral sulfonylureas in patients with diabetes due to Kir6.2 mutations. *New Engl. J. Med.* **355,** 467–477.

Perucca, E., French, J., and Bialer, M. (2007). Development of new antiepileptic drugs: Challenges, incentives, and recent advances. *Lancet Neurol.* **6,** 793–804.

Picard, F., Bertrand, S., Steinlein, O. K., and Bertrand, D. (1999). Mutated nicotinic receptors responsible for autosomal dominant nocturnal frontal lobe epilepsy are more sensitive to carbamazepine. *Epilepsia* **40,** 1198–1209.

Picollo, A., and Pusch, M. (2005). Chloride/proton antiporter activity of mammalian CLC proteins ClC-4 and ClC-5. *Nature* **436,** 420–423.

Picollo, A., Liantonio, A., Didonna, M. P., Elia, L., Conte Camerino, D., and Pusch, M. (2004). Molecular determinants of differential pore blocking of kidney CLC-K chloride channels. *EMBO Rep.* **5,** 584–589.

Picollo, A., Liantonio, A., Babini, E., Conte Camerino, D., and Pusch, M. (2007). Mechanism of interaction of niflumic acid with heterologously expressed kidney CLC-K chloride channels. *J. Membr. Biol.* **216,** 73–82.

Piddlesden, S. J., Jiang, S., Levin, J. L., Vincent, A., and Morgan, B. P. (1996). Soluble complement receptor 1 (sCR1) protects against experimental autoimmune myasthenia gravis. *J. Neuroimmunol.* **71,** 173–177.

Pierno, S., Didonna, M. P., Cippone, V., De Luca, A., Pisoni, M., Frigeri, A., Nicchia, G. P., Svelto, M., Chiesa, G., Sirtori, C., Scanziani, E., Rizzo, C., *et al.* (2006). Effects of chronic treatment with stains and fenofibrate on rat skeletal muscle: A biochemical, histological and electrophysiological study. *Br. J. Pharmacol.* **149,** 909–919.

Pierno, S., De Luca, A., Desaphy, J. F., Fraysse, B., Liantonio, A., Didonna, M. P., Lograno, M., Cocchi, D., Smith, R. G., and Conte Camerino, D. (2003). Growth hormone secretagogues modulate the electrical and contractile properties of rat skeletal muscle through a ghrelin-specific receptor. *Br. J. Pharmacol.* **139,** 575–584.

Pierno, S., Didonna, M. P., Cippone, V., De Luca, A., Pisoni, M., Frigeri, A., Nicchia, G. P., Svelto, M., Chiesa, G., Sirtori, C., *et al.* (2006). Effects of chronic treatment with statins and fenofibrate on rat skeletal muscle: A biochemical, histological and electrophysiological study. *Br. J. Pharmacol.* **149,** 909–919.

Pierno, S., Camerino, G. M., Liantonio, A., Cippone, V., De Luca, A., Giannuzzi, V., Kunic, J., George, A. L., and Conte Camerino, D. (2007). Molecular mechanisms responsible for fluvastatin and fenofibrate-induced reduction of resting membrane chloride conductance in rat skeletal muscle. *Biophys. J.* **92,** 274a (abstract).

Piper, S. N., Triem, J. G., Maleck, W. H., Fent, M. T., Hüttner, I., and Boldt, J. (2001). Placebo-controlled comparison of dolasetron and metoclopramide in preventing postoperative nausea and vomiting in patients undergoing hysterectomy. *Eur. J. Anaesthesiol.* **18,** 251–256.

Priest, B. T., and Kaczorowski, G. J. (2007). Blocking sodium channels to treat neuropathic pain. *Exp. Opin. Ther. Targets* **11,** 291–306.

Priori, S. G., Napolitano, C., Schwartz, P. J., Bloise, R., Crotti, L., and Ronchetti, E. (2000). The elusive link between LQT3 and Brugada syndrome: The role of flecainide challenge. *Circulation* **102,** 945–947.

Proesmans, M., Vermeulen, F., and De Boeck, K. (2008). What's new in cystic fibrosis? From treating symptoms to correction of the basic defect. *Eur. J. Pediatr.* **167,** 839.

Pusch, M. (2002). Myotonia caused by mutations in the muscle chloride channel gene ClCN1. *Hum. Mutat.* **19,** 423–434.

Pusch, M., Liantonio, A., Bertorello, L., Accardi, A., De Luca, A., Pierno, S., Tortorella, V., and Camerino, D. C. (2000). Pharmacological characterization of chloride channels belonging to the ClC family by the use of chiral clofibric acid derivatives. *Mol. Pharmacol.* **58,** 498–507.

Pusch, M., Accardi, A., Liantonio, A., Ferrera, L., De Luca, A., Camerino, D. C., and Conti, F. (2001). Mechanism of block of single protopores of the Torpedo chloride channel ClC-0 by 2-(*p*-chlorophenoxy)butyric acid (CPB). *J. Gen. Physiol.* **118,** 45–62.

Quastoff, S., Spuler, A., Spittelmeister, W., Lehmann-Horn, F., and Grafe, P. (1990). K^+ channel openers suppress myotonic activity of human skeletal muscle *in vitro*. *Eur. J. Pharmacol.* **186,** 125–128.

Ragsdale, D. S. (2008). How do mutant Nav1.1 sodium channels cause epilepsy? *Brain Res. Rev.* **58,** 149–159.

Ragsdale, D. S., McPhee, J. C., Scheuer, T., and Catterall, W. A. (1996). Common molecular determinants of local anesthetic, antiarrhythmic, and anticonvulsant block of voltage-gated Na^+ channels. *Proc. Natl Acad. Sci. USA* **93,** 9270–9275.

Rajakulendran, S., Schorge, S., Kullmann, D. M., and Hanna, M. G. (2007). Episodic ataxia type 1: A neuronal potassium channelopathy. *Neurotherapeutics* **4,** 258–266.

Ramírez, A., Faupel, J., Goebel, I., Stiller, A., Beyer, S., Stöckle, C., Hasan, C., Bode, U., Kornak, U., and Kubisch, C. (2004). Identification of a novel mutation in the coding region of the grey-lethal gene OSTM1 in human malignant infantile osteopetrosis. *Hum. Mutat.* **23,** 471–476.

Reinalter, S. C., Jeck, N., Brochhausen, C., Watzer, B., Nusing, R. M., Seyberth, H. W., and Komhoff, M. (2002). Role of cyclooxygenase-2 in hyperprostaglandin E syndrome/antenatal Bartter syndrome. *Kidney Int.* **62,** 253–260.

Rodgers, H. C., and Knox, A. J. (1999). The effect of topical benzamil and amiloride on nasal potential difference in cystic fibrosis. *Eur. Respir. J.* **14,** 693–696.

Rolland, J. F., Tricarico, D., Laghezza, A., Loiodice, F., Tortorella, V., and Camerino, D. C. (2006). A new benzoxazine compound blocks KATP channels in pancreatic beta cells: Molecular basis for tissue selectivity in vitro and hypoglycaemic action *in vivo*. *Br. J. Pharmacol.* **149,** 870–879.

Rosenbohm, A., Rüdel, R., and Fahlke, C. (1999). Regulation of the human skeletal muscle chloride channel hClC-1 by protein kinase C. *J. Physiol.* **514,** 677–685.

Rowe, S. M., Miller, S., and Sorscher, E. J. (2005). Cystic fibrosis. *N. Engl. J. Med.* **352,** 1992–2001.

Ruan, Y., Liu, N., Bloise, R., Napolitano, C., and Priori, S. G. (2007). Gating properties of SCN5A mutations and the response to mexiletine in long-QT syndrome type 3 patients. *Circulation* **116,** 1137–1144.

Sanders, D. B., El-Salem, K., Massey, J. M., McConville, J., and Vincent, A. (2003). Clinical aspects of MuSK antibody positive seronegative MG. *Neurology* **60,** 1978–1980.

Sanguinetti, M. C., and Tristani-Firouzi, M. (2006). hERG potassium channels and cardiac arrhythmia. *Nature* **440,** 463–469.

Sansone, V., and Tawil, R. (2007). Management and treatment of Andersen–Tawil syndrome (ATS). *Neurotherapeutics* **4,** 233–237.

Schaller, S., Henriksen, K., Sveigaard, C., Heegaard, A. M., Hélix, N., Stahlhut, M., Ovejero, M. C., Johansen, J. V., Solberg, H., Andersen, T. L., *et al.* (2004). The chloride channel inhibitor NS3736 prevents bone resorption in ovariectomized rats without changing bone formation. *J. Bone Miner. Res.* **19,** 1144–1153.

Scheel, O., Zdebik, A. A., Lourdel, S., and Jentsch, T. J. (2005). Voltage-dependent electrogenic chloride/proton exchange by endosomal CLC proteins. *Nature* **436,** 424–427.

Schlingmann, K. P., Weber, S., Peters, M., Niemann Nejsum, L., Vitzthum, H., Klingel, K., Kratz, M., Haddad, E., Ristoff, E., Dinour, D., Syrrou, M., Nielsen, S., *et al.* (2002). Hypomagnesemia with secondary hypocalcemia is caused by mutations in TRPM6, a new member of the TRPM gene family. *Nat. Genet.* **31,** 166–170.

Schneider, J. S., Van Velson, M., Menzaghi, F., and Lloyd, G. K. (1998). Effects of the nicotinic acetylcholine receptor agonist SIB-1508Y on object retrieval performance in MPTP-treated monkeys: Comparison with levodopa treatment. *Ann. Neurol.* **43,** 311–317.

Schneider-Gold, C., Gajados, P., Toyka, K. V., and Hohlfeld, R. R. (2005). Corticosteroids for myasthenia gravis. *Cochrane Database Syst. Rev.* **2, CD002828.**

Schwartz, P. J., Priori, S. G., Locati, E. H., Napolitano, C., Cantù, F., Towbin, J. A., Keating, M. T., Hammoude, H., Brown, A. M., and Chen, L. S. (1995). Long QT syndrome patients with mutations of the SCN5A and HERG genes have differential responses to Na^+ channel blockade and to increases in heart rate: Implications for gene-specific therapy. *Circulation* **92,** 3381–3386.

Sheets, P. L., Jackson, J. O., II, Waxman, S. G., Dib-Hajj, S. D., and Cummins, T. R. (2007). A $Na_v1.7$ channel mutation associated with hereditary erythromelalgia contributes to neuronal hyperexcitability and displays reduced lidocaine sensitivity. *J. Physiol.* **581**(3), 1019–1031.

Shillito, P., Molenaar, P. C., Vincent, A., Leys, K., Zheng, W., van den Berg, R. J., Plomp, J. J., van Kempen, G. T. H., Chauplannaz, G., Wintzen, A. R., van Dijk, J. G., and Newsom-Davis, J. (1995). Acquired neuromyotonia: Evidence for autoantibodies directed against K+ channels of peripheral nerves. *Ann. Neurol.* **38,** 714–722.
Shimizu, W., Aiba, T., and Antzelevitch, C. (2005). Specific therapy based on the genotype and cellular mechanism in inherited cardiac arrhythmias Long QT syndrome and Brugada syndrome. *Curr. Pharm. Des.* **11,** 1561–1572.
Silberstein, S. D. (2006). Preventive treatment of migraine. *Trends Pharmacol. Sci.* **27,** 410–415.
Simard, J. M., Woo, K. S., Bhatta, S., and Gerzanich, V. (2008). Drugs acting on SUR1 to treat CNS ischemia and trauma. *Curr. Opin. Pharmacol.* **8,** 42–49.
Singer, W., Opfer-Gehrking, T. L., McPhee, B. R., Hilz, M. J., Bharucha, A. E., and Low, P. A. (2003). Acetylcholinesterase inhibition: A novel approach in the treatment of neurogenic orthostatic hypotension. *J. Neurol. Neurosurg. Psychiatry* **74,** 1294–1298.
Snyder, P. M., Price, M. P., McDonald, F. J., Adams, C. M., Volk, K. A., Zeiher, B. G., Stokes, J. B., and Welsh, M. J. (1995). Mechanism by which Liddle's syndrome mutations increase activity of a human epithelial Na^+ channel. *Cell* **83,** 969–978.
Soreq, H., and Seidman, S. (2000). Anti-sense approach to anticholinesterase therapeutics. *Isr. Med. Assoc. J.* **2**(Suppl.), 81–85.
Steinlein, O. K. (2004). Genetic mechanisms that underlie epilepsy. *Nat. Rev. Neurosci.* **5,** 400–408.
Steinmeyer, K., Ortland, C., and Jentsch, T. J. (1991). Primary structure and functional expression of a developmentally regulated skeletal muscle chloride channel. *Nature* **354,** 301–304.
Striano, P., Coppola, A., Pezzella, M., Ciampa, C., Specchio, N., Ragona, F., Mancarde, M. M., Gennaro, E., Beccaria, F., Capovilla, G., Rasmini, P., Befana, D., *et al.* (2007). An open-label trial of levetiracetam in severe myoclonic epilepsy of infancy. *Neurology* **69,** 250–254.
Strupp, M., Zwergal, A., and Brandt, T. (2007). Episodic ataxia type 2. *Neurotherapeutics* **4,** 267–275.
Takahashi, M. P., and Cannon, S. C. (2001). Mexiletine block of disease-associated mutations in S6 segments of the human skeletal muscle Na^+ channel. *J. Physiol.* **537,** 701–714.
Tate, S. K., Depondt, C., Sisodiya, S. M., Cavalleri, G. L., Schorge, S., Soranzo, N., Thom, M., Sen, A., Shorvon, S. D., Sander, J. W., Wood, N. W., and Goldstein, D. B (2005). Genetic predictors of the maximum doses patients receive during clinical use of the anti-epileptic drugs carbamazepine and phenytoin. *Proc. Natl. Acad. Sci. USA* **102,** 5507–5512.
Tawil, R., McDermott, M. P., Brown, R., Jr., Shapiro, B. C., Ptacek, L. J., McManis, P. G., Dalakas, M. C., Spector, S. A., Mendell, J. R., Hahn, A. F., and Griggs, R. C. (2000). Randomized trials of dichlorphenamide in the periodic paralyses. Working Group on Periodic Paralysis. *Ann. Neurol.* **47,** 46–53.
Teulon, J., Lourdel, S., Nissant, A., Paulais, M., Guinamard, R., Marvao, P., and Imbert-Teboul, M. (2005). Exploration of the basolateral chloride channels in the renal tubule using. *Nephron Physiol.* **99,** 64–68.
Thiry, A., Dogné, J. M., Supuran, C. T., and Marercel, B. (2008). Anticonvulsant sulfonamides/sulfamates/sulfamides with carbonic anhydrase inhibitory activity: Drug design and mechanism of action. *Curr. Pharm. Des.* **14,** 661–671.
Timerman, A. P., Ogunbumni, E., Freund, E., Wiederrecht, G., Marks, A. R., and Fleischer, S. (1993). The calcium release channel of sarcoplasmic reticulum is modulated by FK-506-binding protein. Dissociation and reconstitution of FKBP-12 to the calcium release channel of skeletal muscle sarcoplasmic reticulum. *J. Biol. Chem.* **268,** 22992–22999.
Tricarico, D., Pierno, S., Mallamaci, R., Brigiani, G. S., Santoro, G., and Camerino, D. C. (1998). The biophysical and pharmacological characteristics of skeletal muscle KATP channels are modified in K^+ depleted rat, an animal model of hypokalemic periodic paralysis. *Mol. Pharmacol.* **54,** 197–206.
Tricarico, D., Servidei, S., Tonali, P., Jurkat-Rott, K., and Conte Camerino, D. (1999). Impairment of skeletal muscle adenosine triphosphate-sensitive K^+ channels in patients with hypokalemic periodic paralysis. *J. Clin. Invest.* **103,** 675–682.

Tricarico, D., Barbieri, M., Antonio, L., Tortorella, P., Loiodice, F., and Conte Camerino, D. (2003). Dualistic actions of cromakalim and new potent 2H-1,4-benzoxazine derivatives on the native skeletal muscle KATP channel. *Br. J. Pharmacol.* **139,** 255–262.

Tricarico, D., Barbieri, M., Mele, A., Carbonara, G., and Camerino, D. C. (2004). Carbonic anhydrase inhibitors are specific openers of skeletal muscle BK channel of K^+-deficient rats. *FASEB J.* **18,** 760–761.

Tricarico, D., Mele, A., and Conte Camerino, D. (2006a). Carbonic anhydrase inhibitors ameliorate the symptoms of hypokalaemic periodic paralysis in rats by opening the muscular Ca^{2+}-activated-K^+ channels. *Neuromuscul. Disord.* **16,** 39–45.

Tricarico, D., Mele, A., Lundquist, A. L., Desai, R. R., George, A. L., Jr., and Conte Camerino, D. (2006b). Hybrid assemblies of ATP-sensitive K^+ channels determine their muscle type-dependent biophysical and pharmacological properties. *Proc. Natl Acad. Sci. USA* **103,** 1118–1123.

Tricarico, D., Lovaglio, S., Mele, A., Rotondo, G., Mancinelli, E., Meola, G., and Conte Camerino, D. (2008a). Acetazolamide prevents vacuolar myopathy in skeletal muscle of K-depleted rats. *Br. J. Pharmacol.* **154,** 183–190.

Tricarico, D., Mele, A., Liss, B., Ashcroft, F. M., Lundquist, A. L., Desai, R. R., George, A. L., Jr., and Conte Camerino, D. (2008b). Reduced expression of Kir6.2/SUR2A subunits explains KATP deficiency in K^+-depleted rats. *Neuromuscul. Disord.* **18**(1), 74–80.

Trip, J., Drost, G., van Engelen, B. G., and Faber, C. G. (2006). Drug treatment for myotonia. *Cochrane Database Syst. Rev.* **25, CD004762.**

Trudeau, M. M., Dalton, J. C., Day, J. W., Ranum, L. P., and Meisler, M. H. (2006). Heterozygosity for a protein truncation mutation of sodium channel SCN8A in a patient with cerebellar atrophy, ataxia, and mental retardation. *J. Med. Genet.* **43,** 527–530.

Ubiali, F., Nava, S., Nessi, V., Longhi, R., Pezzoni, G., Capobianco, R., Mantegazza, R., Antozzi, C., and Baggi, F. (2008). Pixantrone (BBR2778) reduces the severity of experimental autoimmune myasthenia gravis in Lewis rats. *J. Immunol.* **180,** 2696–2703.

Unwin, R. J., and Capasso, G. (2006). Bartter's and Gitelman's syndromes: Their relationship to the actions of loop and thiazide diuretics. *Curr. Opin. Pharmacol.* **6,** 208–213.

Venance, S. L., Cannon, S. C., Fialho, D., Fontaine, B., Hanna, M. G., Ptacek, L. J., Tristani-Firouzi, M., Tawil, R., and Griggs, R. C., and CINCH Investigators (2006). The primary periodic paralyses: Diagnosis, pathogenesis and treatment. *Brain* **129,** 8–17.

Venance, S. L., Herr, B. E., and Griggs, R. C. (2007). Challenges in the design and conduct of therapeutic trials in channel disorders. *Neurotherapeutics* **4,** 199–204.

Vernino, S. (2007). Autoimmune and paraneoplastic channelopathies. *Neurotherapeutics* **4,** 305–314.

Vernino, S., Sandroni, P., Singer, W., and Low, P. A. (2008). Invited article: Autonomic ganglia: Target and novel therapeutic tool. *Neurology* **70,** 1926–1932.

Vidic-Dankovic, B., Kosec, D., Damjanovic, M., Apostolski, S., Isakovic, K., and Bartlett, R. R. (1995). Leflunomide prevents the development of experimentally induced myasthenia gravis. *Int. J. Immunopharmacol.* **17,** 273–281.

Vollmer, T., Key, L., Durkalski, V., Tyor, W., Corboy, J., Markovic-Plese, S., Preiningerova, J., Rizzo, M., and Singh, I. (2004). Oral simvastatin treatment in relapsing-remitting multiple sclerosis. *Lancet* **363,** 1607–1608.

Walder, R. Y., Landau, D., Meyer, P., Shalev, H., Tsolia, M., Borochowitz, Z., Boettger, M. B., Beck, G. E., Englehardt, R. K., Carmi, R., and Sheffield, V. C. (2002). Mutation of TRPM6 causes familial hypomagnesemia with secondary hypocalcemia. *Nat Genet.* **31,** 171–174.

Walz, W. (2002). Chloride/anion channels in glial cell membranes. *Glia* **40,** 1–10.

Wang, D. W., Yazawa, K., Makita, N., George, A. L., Jr., and Bennett, P. B. (1997). Pharmacological targeting of long QT mutant sodium channels. *J. Clin. Invest.* **99,** 1714–1720.

Wang, L., Kinnear, C., Hammel, J. M., Zhu, W., Hua, Z., Mi, W., and Caldarone, C. A. (2006). Preservation of mitochondrial structure and function after cardioplegic arrest in the neonate using a selective mitochondrial KATP channel opener. *Ann. Thorac. Surg.* **81,** 1817–1823.

Watanabe, H., Koopmann, T. T., Le Scouarnec, S., Yang, T., Ingram, C. R., Schott, J. J., Demolombe, S., Probst, V., Anselme, F., Escande, D., *et al.* (2008). Sodium channel beta1 subunit mutations associated with Brugada syndrome and cardiac conduction disease in humans. *J. Clin. Invest.* **118,** 2260–2268.

Webb, T. I., and Lynch, J. W. (2007). Molecular pharmacology of the glycine receptor chloride channel. *Curr. Pharm. Des.* **13,** 2350–2367.

Weinstock-Guttman, B., Ramanathan, M., Lincoff, N., Napoli, S. Q., Sharma, J., Feichter, J., and Bakshi, R. (2006). Study of mitoxantrone for the treatment of recurrent neuromyelitis optica (Devic disease). *Arch. Neurol.* **63,** 957–963.

Winn, M. P., Conlon, P. J., Lynn, K. L., Farrington, M. K., Creazzo, T., Hawkins, A. F., Daskalakis, N., Kwan, S. Y., Ebersviller, S., Burchette, J. L., Pericak-Vance, M. A., Howell, D. N., *et al.* (2005). A mutation in the TRPC6 cation channel causes familial focal segmental glomerulosclerosis. *Science* **308,** 1801–1804.

Wintzen, A. R., Lammers, J. G., and van Dijk, J. G. (2007). Does modafinil enhance activity of patients with myotonic dystrophy? A double-blind placebo-controlled crossover study. *J. Neurol.* **254,** 26–28.

Wu, L., Shen, F., Lin, L., Zhang, X., Bruce, C. I., and Xia, Q. (2006). The neuroprotection conferred by activating the mitochondrial ATP-sensitive K^+ channel is mediated by inhibiting the mitochondrial permeability transition pore. *Neurosci. Lett.* **402,** 184–189.

Yadava, R. S., Frenzel-McCardell, C. D., Yu, Q., Srinivasan, V., Tucker, A. L., Puymirat, J., Thornton, C. A., Prall, O. W., Harvey, R. P., and Mahadevan, M. S. (2008). RNA toxicity in myotonic muscular dystrophy induces NKX2-5 expression. *Nat. Genet.* **40,** 61–68.

Yarov-Yarovoy, V., McPhee, J. C., Idswoog, D., Pate, C., Scheuer, T., and Catterall, W. A. (2002). Role of amino acid residues in transmembrane segments IS6 and IIS6 of the Na channel α-subunit in voltage-dependent gating and drug block. *J. Biol. Chem.* **277,** 35393–35401.

Zaja, F., Russo, D., Fuga, G., Perella, G., and Baccarani, M. (2000). Rituximab for myasthenia gravis developing after bone marrow transplant. *Neurology* **55,** 1062–1063.

Zeitlin, P. L., Boyle, M. P., Guggino, W. B., and Molina, L. (2004). A phase I trial of intranasal Moli1901 for cystic fibrosis. *Chest* **125,** 143–149.

Zeng, H., Lozinskaya, I. M., Lin, Z., Willette, R. N., Brooks, D. P., and Xu, X. (2006). Mallotoxin is a novel human ether-a-go-go-related gene (hERG) potassium channel activator. *J. Pharmacol. Exp. Ther.* **319,** 957–962.

Zhou, L., Chillag, K. L., and Nigro, M. A. (2002). Hyperekplexia: A treatable neurogenetic disease. *Brain Dev.* **24,** 669–674.

Zipes, D. P., Camm, A. J., Borggrefe, M., Buxton, A. E., Chaitman, B., Fromer, M., Gregoratos, G., Klein, G., Moss, A. J., Myerburg, R. J., Priori, S. G., Quinones, M. A., *et al.* (2006). ACC/AHA/ESC 2006 Guidelines for management of patients with ventricular arrhythmias and the prevention of sudden cardiac death-Executive summary. *Circulation* **114,** 1088–1132.

Zünkler, B. J. (2006). Human ether-a-go-go-related (HERG) gene and ATP-sensitive potassium channels as targets for adverse drug effects. *Pharmacol. Ther.* **112,** 12–37.

Index